T0289254

Object and Pattern Recognition in Remote Sensing

Modelling and Monitoring Enviromental and Anthropogenic Objects and Change Processes

Edited by
S. Hinz, A.C. Braun, M. Weinmann

WHITTLES PUBLISHING

Published by
Whittles Publishing,
Dunbeath,
Caithness KW6 6EG,
Scotland, UK

www.whittlespublishing.com

© 2021 S. Hinz, A.C. Braun, M. Weinmann

ISBN 978-184995-128-9

All rights reserved.
No part of this publication may be reproduced, stored in a retrieval system, or transmitted, in any form or by any means, electronic, mechanical, recording or otherwise without prior permission of the publishers.

The publisher and authors have used their best efforts in preparing this book, but assume no responsibility for any injury and/or damage to persons or property from the use or implementation of any methods, instructions, ideas or materials contained within this book. All operations should be undertaken in accordance with existing legislation, recognized codes and standards and trade practice. Whilst the information and advice in this book is believed to be true and accurate at the time of going to press, the authors and publisher accept no legal responsibility or liability for errors or omissions that may have been made.

Printed in Malta by Melita Press

Contents

Foreword

An image says more than a thousand words – this well-known proverb, which has its origin in the advertisement industry, describes much of the fascination of photogrammetry and remote sensing, and in particular of its images. In sharp contrast to machines – and the statement remains true also in today's era of deep learning and artificial intelligence – humans are capable of interpreting images and videos, and of extracting the information contained therein without much effort and in real-time. However, we do not know just how we accomplish this task, which perhaps explains some of the fascination of aerial and satellite imagery.

Major characteristics of photogrammetry and remote sensing are:

- contact-free data acquisition,
- short image acquisition time which allows the capture of dynamic processes,
- derivation of 3D information from stereoscopic images,
- complete iconic documentation of the whole scene,
- the possibility to measure and monitor objects of arbitrary size; methods in use vary in scale from microscopy all the way to planetary remote sensing.

Traditionally, remote sensing deals with earth observation from space – monitoring of processes on the Earth's surface, as well as in the atmosphere are of prime interest – while the focus of photogrammetry is the 3D geometric and thematic exploitation of images of all kind. Technical innovations, however, have led to more and more overlap between the two disciplines. Photogrammetry was of course established long before satellites became available; nevertheless it is today regarded as part of the wider field of remote sensing. Moreover, photogrammetry forms one of the foundations of modern computer vision.

In recent years, photogrammetry and remote sensing have witnessed great changes in virtually every stage of image acquisition and processing. These developments have had a profound impact on the theory, development and operational use of photogrammetry and remote sensing. Examples include:

- new and better sensors and platforms such as 3D cameras, flash lidars, unmanned aerial vehicles and mobile mapping vehicles,
- the deployment of satellite constellations with the aim of imaging the earth in high resolution every single day,
- a shift from single sensors to integrated sensor systems, e. g. for assisted and autonomous driving,

- an increasing use of, and integration with, methods developed in computer science and computer vision, both for geometric and sematic tasks,
- a shift from mapping to monitoring and to dynamic geospatial services, e.g. for sustainable development.

Innovation in our field has been strongly influenced by progress in information and communication technology, and terms like ubiquitous computing, geosensor networks, digital earth, big data, deep learning, cloud computing, the semantic web, the internet of things and crowd sourcing have become part of our discipline. Besides a large variety of new applications, all the way from robotics and driver-assistance systems to animal behaviour studies and environmental monitoring, new methodologies for automatic, reliable information extraction and cooperative decision making employing concepts from data mining and machine learning, are now commonplace in our field.

This observation leads directly to the title of this book – *Object and pattern recognition in remote sensing.* The book reports on research and development projects which were carried out in approximately the last 10 years at the Institute of Photogrammetry and Remote Sensing of the Karlsruhe Institute of Technology (KIT) under the leadership of Professor Stefan Hinz, and it gives an excellent overview of the current state-of-the-art in photogrammetry and remote sensing. For decades KIT has been one of the prime centres in this area, and the work documented in this book proves that this remains the case. While the book is not designed as a text book, it is of high relevance to students and other people wanting to learn about photogrammetry and remote sensing. The individual chapters discuss a number of important topics and, at the same time, show the clear direction of the current activities, at KIT and elsewhere, in photogrammetry and remote sensing research.

I believe such success has basically two ingredients: (a) individual excellence, which the KIT group greatly possesses, and (b) an international network, to be able to share and learn from latest trends in their own discipline and in neighbouring fields. Together with his group, Stefan Hinz is very active in the network set up and maintained within the International Society of Photogrammetry and Remote Sensing (ISPRS).

ISPRS, www.isprs.org, is an international non-governmental organization that promotes international cooperation between the worldwide organizations with interests in the photogrammetry, remote sensing and spatial information sciences. Established in 1910, ISPRS is the oldest international umbrella organization in its field, which may be summarized as addressing "information from imagery". The ISPRS scientific and technical programmes are organized by five Commissions. Each Commission is sponsored by an ISPRS Member organization for the four-year period between Congresses. The Commissions have established more than 60 Working Groups which are responsible for particular topics within the Commissions' areas of responsibility.

KIT has been active in ISPRS for a very long time. As early as 1964 Kurt Schwidefsky, one of the predecessors of Stefan Hinz and later an ISPRS Honorary Member, became President of the ISPRS Commission II, then entitled "Theory, methods, instruments of restitution". From 2016 – 2020, Stefan Hinz is President of Commission I, "Sensor systems"; in the preceding period Boris Jutzi held the position of Vice President of the same Commission, continuing in this position for the present period. In addition, scientists from KIT chaired ISPRS working groups, organised ISPRS events and presented numerous papers at the ISPRS Congresses, symposia and workshops. As mentioned, such engagement is an essential ingredient in modern science and development, which in the days of open science, open source, open data and open access is heavily influenced by international and interdisciplinary cooperation.

I congratulate the authors, and in particular the editors under the leadership of Stefan Hinz, for this nice piece of work and hope that the readers will find the material as interesting as I did.

Christian Heipke
ISPRS President 2016–2020
Institut für Photogrammetrie und GeoInformation (IPI)
Leibniz Universität Hannover

Editors and Authors

Stefan Hinz studied geodesy and geoinformatics at the TU Muenchen (MSc, 1998) and received the doctorate in 2003 and the *venia legendi* (Habilitation) in 2008. His research work concentrated on image analysis for automatic mapping from aerial and satellite images by methods of artificial intelligence. In the course of the habilitation, this research extended towards thermal and radar data with application to short and long-term monitoring of moving objects, land use and urban infrastructure. With the appointment in 2008 as full professor for Remote Sensing and Image Processing at KIT, Professor Hinz became director of the Institute for Photogrammetry and Remote Sensing (IPF) and has since then been leading a group of scientists of about 30–35 people. He is head of the Graduate School for Climate and Environment (GRACE) of approximately 100 PhD students and served as dean of the Faculty for Civil Engineering, Geo- und Environmental Sciences for the period 2012–2018. Since 2016, he has been President of Technical Commission 1 *Sensors and Platforms* of the International Society for Photogrammetry and Remote Sensing (ISPRS) and member of the ISPRS Journal editorial advisory board.

Andreas Braun received the Diploma in Geoecology from Technical University of Karlsruhe in 2009. After continuing his research in remote sensing at Karlsruhe Institute of Technology (Institute of Photogrammetry and Remote Sensing), he received his PhD in 2013 and spent a year as a postdoctoral researcher at the university of Freiburg (Chair of Remote Sensing and Landscape Information Systems). From 2011 he additionally started his studies of sociology, economy and philosophy at the FernUniversität in Hagen (Germany) and received his MA in sociology in 2016. In 2015 he was assigned a junior professorship for risk oriented regional development at the Institute of Regional Science (KIT), which he currently holds, investigating human vulnerability in developing countries (mainly in Latin America).

Martin Weinmann received the Diploma in Electrical Engineering and Information Technology from the Technical University of Karlsruhe in 2009 and the PhD degree from the Karlsruhe Institute of Technology (KIT) in 2015. Currently, he is postdoctoral researcher at the Institute of Photogrammetry and Remote Sensing at KIT. His main research interests are in the fields of computer vision, pattern recognition, active 3D vision, image and point cloud processing, and scene analysis. He served several years as reviewer for the International Society for Photogrammetry and Remote Sensing (ISPRS), the Institute of Electrical and Electronics Engineers (IEEE) and the German Society for Photogrammetry, Remote Sensing and Geoinformation (DGPF).

Callum Banfield received his Diploma in Geoecology from the Karlsruhe Institute of Technology in 2013. After having spent one year travelling, he started pursuing a PhD in soil science at the Department of Temperate Ecosystems of the Georg-August University of Goettingen. During his PhD studies he is focusing on subsoil carbon dynamics of arable fields, combining biomarker approaches with stable and radioactive tracers.

Clémence Dubois received her Double-Diploma in Topography, Geodesy and Geomatics from the National Institute of Applied Sciences (INSA) of Strasbourg, France, and the Karlsruhe Institute of Technology (KIT), Germany, in 2011, and her PhD degree from the Karlsruhe Institute of Technology (KIT) in 2015. From 2012 to mid-2016, she has been with the Institute of Photogrammetry and Remote Sensing (IPF) of the Karlsruhe Institute of Technology (KIT), and also a member of the GRAduate School for Climate and Environment (GRACE) of KIT. Since 2016, she has been with the department "Geoscientific Information,

International Cooperation" of the Federal Institute for Geosciences and Natural Resources, Germany. Her research activity mainly covers SAR, InSAR and stereo-SAR image exploitation, especially object detection and reconstruction, change detection, and lately DInSAR and PS processing for diverse geo-hazard and remote sensing applications.

Franz Rottensteiner received a Dipl. Ing. Degree in surveying and PhD and venia docendi in photogrammetry, all from Vienna University of Technology, Vienna, Austria. He has been a Postdoctoral Researcher with the Vienna University of Technology, the University of New South Wales (Kensington, NSW, Australia), and the University of Melbourne (Parkville, Victoria, Australia). He is currently with the Institute of Photogrammetry and GeoInformation, Leibniz Universität Hannover (Hannover, Germany), where he is an Associate (apl.) Professor, leading the Photogrammetric Image Analysis Group. His research interests include photogrammetry, automated extraction of topographic objects, processing of lidar data, and sensor orientation.

Miriam Hommel received her Diploma in Geodesy and Geoinformatics from Universität Karlsruhe (TH) in 2006 and her PhD degree from Karlsruhe Institute of Technology (KIT) in 2010. From 2006 to 2010 she was staff member of the Institute of Photogrammetry and Remote Sensing (IPF) at KIT and from 2010 to 2011 of the Fraunhofer Institute for Manufacturing Engineering and Automation (IPA). Since 2012 she has been an employee at Aalen University of Applied Sciences. Her main research interests are object detection and reconstruction from lidar and multi-spectral data, automatic change and damage detection, as well as classification and statistical data analysis.

Boris Jutzi received a Diploma in Electrical Engineering at University of Kaiserslautern, Germany, obtained his PhD degree at Technische Universität München (TUM), Germany, and the Venia Legendi at Karlsruhe Institute of Technology (KIT), Germany. Currently he is Adjunct Professor and head of the "Active Sensors & Computer Vision" Department at the Institute of Photogrammetry and Remote Sensing at KIT. His main research interests are computer vision, lidar remote sensing, automatic image analysis, and optical active sensing with full-waveform lidar and time-of-flight imaging. He served several years for the International Society for Photogrammetry and Remote Sensing (ISPRS), e.g. 2012–2016 as Vice President of the ISPRS Technical Commission I "Sensors and Platforms for Remote Sensing" and 2016–2020 as Secretary of the ISPRS Technical Commission I "Sensor Systems".

Sina Keller holds a degree in mathematics and geography from the University of Karlsruhe in 2010. She received her PhD degree from the Karlsruhe Institute of Technology (KIT) in 2015. During her PhD she worked on a model assessing natural hazards on critical infrastructure. Currently, she is a postdoctoral researcher at the Institute of Photogrammetry and Remote Sensing at KIT. Since 2012, she has been working as a lecturer for mathematics at the Baden-Wuerttemberg Cooperative State University Karlsruhe. In 2015, she also took over leadership of the research division at the company ci-Tec GmbH. Her main research interests address the modelling of hydrological processes with hyperspectral and thermal data as well as the modelling and investigation of road network vulnerability.

Jens Leitloff received his Diploma in Geodetic Engineering from the Technische Universität Berlin in 2003 and his PhD degree from Technische Universität München in 2011. From 2008 to 2011 he was with the Remote Sensing Technology Institute at the German Aerospace Center (DLR) in Oberpfaffenhofen. In 2011 he joined the Institute of Photogrammetry and Remote Sensing at Karlsruhe Institute of Technology (KIT). His main research interests are computer vision, image processing, image and point cloud classification, camera systems and camera calibration.

Florian Schmidt received his Diploma in Geodesy and Geoinformatics from the Technical University of Karlsruhe in 2009 and his PhD degree from the Karlsruhe Institute of Technology (KIT) in 2012. From 2009 to 2012, he was staff member of the Institute of Photogrammetry and Remote Sensing (IPF) at KIT. His main research interests include object detection and tracking, machine learning and analysis of aerial image sequences.

Ulrike Sturm-Hentschel (nee Ulrike Sturm) received her diploma in geoecology in 2004 after studying at Technical University in Karlsruhe (today: Karlsruhe Institute of Technology) and the Universidade Federal do Parana in Brazil. From 2004 to 2006 she worked as a scientific volunteer at Staatliches Museum für Naturkunde (Federal Museum of Natural History) in Karlsruhe. From 2006 to 2011 she has been working as a scientific assistant at the Institute of Photogrammetry and Remote Sensing of Karlsruhe Insitute of Technology. In 2012, she received her PhD for her study on the settlement processes in the coastal area of Benin, in which she applied several remote sensing techniques for a spatially explicit assessment of settlement growth. These studies were backed up by field research, and the two types of data sets were related to each other. From 2012 to 2013 she has been working as a freelancer in the field of sustainable development. Since 2013, she has been a scientific assistant at the German Museum in Munich, responsible for the Department of Ground Transportation.

Antje Thiele received her Diploma in Geodesy from the Technical University Dresden in 2004 and her PhD degree from Leibniz Universität Hannover (LUH) in 2013. Since 2005, she has been with the Fraunhofer Institute of Optronics, System Technologies and Image Exploitation (IOSB) and since 2009 also with the Institute of Photogrammetry and Remote Sensing (IPF) at the Karlsruhe Institute of Technology (KIT). In 2014, she took over leadership of the group "SAR Image Exploitation" at Fraunhofer IOSB. Her research activity mainly concerns SAR and InSAR image exploitation by focusing on object detection and reconstruction, classification, change detection and simulation.

Uwe Weidner received his Diploma in Geodesy in 1992 and his PhD degree in 1997, both from the Rheinische Friedrich-Wilhelms University, Bonn. From 1997 to 2000 he worked for a company in the field of aerial photogrammetry and from 2000 to 2002 for a company in the field of industrial image processing and robotics. Since February 2003 he has been with the Institute of Photogrammetry and Remote Sensing, KIT, Karlsruhe, as senior scientist in the field of remote sensing. His research interests include various aspects of automatic object extraction using multi-/hyperspectral remote sensing data and evaluation of methods in remote sensing. He served several years as reviewer for the International Society for Photogrammetry and Remote Sensing (ISPRS), the Institute of Electrical and Electronics Engineers (IEEE) and the German Society for Photogrammetry, Remote Sensing and Geoinformation (DGPF).

Thomas Vögtle received his Diploma in Geodesy in 1982 and his PhD degree in 1989, both from the University of Karlsruhe. Since 1984, he has been a member of the Institute of Photogrammetry and Remote Sensing, working on several scientific projects. Since 1991 he has been Senior Scientist and Lecturer at the same institute. His research activities concern automation in image analysis and point cloud processing as well as 3D object extraction from different data sets. He has served as reviewer for several international journals, has published two lecture books as editor/co-author, and has received two awards (Dambach award, Hansa-Luftbild award).

Sven Wursthorn received his Diploma in Geodesy from the Technical University of Karlsruhe in 2002 and his PhD degree from the Karlsruhe Institute of Technology (KIT) in 2014. Since 2002, he has been with the Institute of Photogrammetry and Remote Sensing (IPF) at the Karlsruhe Institute of Technology (KIT). His main research interests are geographic information systems, augmented reality and photogrammetry.

Part I
Methodology

Introduction

CHAPTER 1

Stefan Hinz

There have been many attempts to grasp the meaning of the main topic of this book – image and scene analysis. Summarising the perceptions of many review books and articles, image and scene analysis focuses on the automated characterisation of a certain part of the real world – the scene – recorded by a specific sensor. In the context of remote sensing (as a discipline), the majority of these devices are imaging sensors, hence relating scene analysis closely to the general tasks of image processing. Indeed, it is often used synonymously with terms like *image understanding* or even *scene interpretation*[1]. A nice historical review of the origins of digital image processing, which can be traced back to the early 1960s, can be found in Gonzalez and Woods (2002). Condensing the various aspects of image and scene analysis into one sentence, Rosenfeld and Kak (1982) stated that image analysis is defined as "the automatic derivation of a meaningful description of physical objects in the real world from images". The aim of image analysis is thus to generate an explicit description of objects from the information implicitly contained in the data. This task can be compared to the human skill of photo interpretation. The primary aim is to achieve the same results of interpretation as a human would achieve, although how a human actually interprets images plays a minor role here. Needless to say, the human approach to image interpretation can nevertheless give useful hints when designing remote sensing image analysis frameworks.

1.1 Sensors and Imaging

When transferring the opportunities and potentials of image analysis to disciplines such as remote sensing, computer vision or photogrammetry, the term *image* adopts a much broader notion than its original sense of a digital frame image (i.e. a regularly sampled array of pixels) taken in the visible spectrum. Modern remote sensing and computer vision not only involve spectra far beyond the visible domain, but also make use of active sensors such as radar and laser scanning sensors (also known as lidar sensors). These kinds of sensor emit specifically coded signals and record the echo(es) of exactly these signals, thereby allowing analysis of a scene and the objects therein with a specific focus on the properties of the emitted signals. A more in-depth discussion of the different sensors and sensor concepts will be given in the next chapter.

Having been underestimated in the very early years of its development, the tasks related to image analysis are known to be very hard to solve, depending on the complexity of the data and the scene to be analysed, in particular when dealing with outdoor environments. In addition to the challenges of scene analysis arising from the complex structure of the world, there are further challenges in recording and storing the results from the process of image formation. The decisive factor in this is the sensor, as well as the characteristics of the information carrier. Usually, electromagnetic radiation (EMR) serves as the information carrier, but other types of waves, in particular acoustic waves, are sometimes used. The information carrier triggers, which characteristics of the objects in the real world can be recorded by the sensor – in the case of EMR, the radiation and reflection properties of the objects. The sensor, meanwhile, determines how this information is effectively recorded and stored. Here, a series of geometric and physical transformations come into play, which change the original information fundamentally. These may include a geometrical transformation from 3D to 2D, the spatial discretisation of the image by the CCD[2] or CMOS[3] sensor elements, physical conversion of the EMR into a measurable physical entity (e.g. charge), and finally its quantisation into bits or integer numbers.

1.2 From Image Information to Semantics of Objects

If we transfer this concept to typical remote sensing data[4], which are most relevant in the context of this book, the following characteristics (regarding the information contained in the image) arise:

- *Radiometric information*: The intensity values are approximately proportional to the reflection properties of the objects, but limited by the discrete radiometric and spectral resolution of the sensor.
- *Geometric information*: The original geometry of the objects now exists only in a discrete representation. In the case of common frame images and assuming a pinhole camera, the dimension is reduced by the central perspective projection from 3D to 2D. This may lead to a change in the topology of adjacent scene elements in certain image areas. In the case of occlusions, for example, two adjacent pixels do not correspond to two adjacent surface elements in the real world. This holds similarly for other imaging geometries in remote sensing, and also for 3D sensors like laserscanners or time-of-flight cameras.
- *Semantic information*: Objects or object classes and their semantics are defined only by human beings; that is, semantics are not attached to the objects in a physical sense, and the meaning of objects is only assigned by humans interpreting the radiometric and geometric information of the images. The recorded data only implicitly contain information about the semantics of the imaged scene. Transforming this implicitly contained information into explicit information is thus the key issue of image and scene analysis.

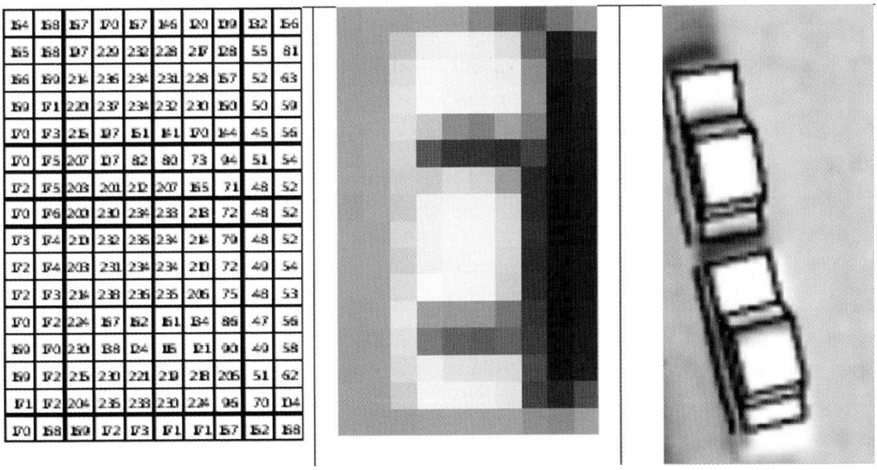

Figure 1.1 Example showing (a) radiometric information represented as digital numbers; (b) the same information represented as greyvalues; (c) assigned semantics in the form of an explicit vehicle model.

Figure 1.1 illustrates the aspects mentioned above. The radiometric information transformed into 8-bit digital numbers ranging from 0 (black) to 255 (white) of a small part of a panchromatic aerial image is shown in Figure 1.1a. Figure 1.1b visualises the same implicit information, but displayed in greyvalues. In contrast, Figure 1.1c shows the result after applying an image analysis approach designed for detecting vehicles (Hinz, 2004). Now, as a vehicle model representing the vehicle as well as its substructures (such as windows, hood and rear) has been fitted to the greyvalues, the semantics of the corresponding greyvalue patches are available; thus, the implicit information has been made explicit.

Determining the semantics of the objects is only possible through the integration of additional knowledge. Usually, one can divide this into two components:

- The *model*, comprising the radiometric, geometric and semantic aspects of the object(s). Knowledge is represented by a model that describes a specification of the general imaging process with respect to the objects to be extracted. The knowledge contained in the model forms the basis for the identification of objects from the raw intensity values.
- The *extraction* strategy, which builds the conceptual basis of the processing scheme to automatically extract the objects from the image, thereby making use of the model. The extraction strategy is the formalism that defines individual image processing operations and the systematics of their application. As the strategy determines the use of the model knowledge, it also has the same relevance as the model itself.

The rationale behind the extraction strategy is very similar to model inversion or parameter estimation schemes (e.g. inversion or search algorithms). It aims to invert the process of image generation and also to determine the object semantics and object relations and their specific properties in the scene from the recorded raw image data. To this end, the utilisation of 3D information turns out to be extremely helpful. Sensors and methods like laser scanning, interferometric synthetic aperture radar (SAR) and time-of-flight imaging, among others, deliver 3D data directly. However, with 2D image data, by using multiple images providing different views 3D information can be reconstructed for static scenes. If the orientation of the images is known – at least theoretically – each object point, that is included in two or more images can be defined in 3D by means of well-known photogrammetric methods (e.g. the advanced approach of Hirschmueller, 2008). This task, in general, is not of a semantic nature and can thus be automated quite reliably.

1.3 Conclusion

Automatic object extraction, in general, thus rests on two fundamentals: an appropriate computer-oriented formalisation of knowledge (modelling) and the use of knowledge for extraction (strategy). For both, a variety of different and complementary approaches exist. An object model, for instance, may rest on a specific, parametric (CAD-like) 3D description or on a view-based description automatically learned from many example images of a particular object (or even learned from simulated images). Approaches for extraction strategies range from more or less monolithic image processing pipelines designed for the extraction of a specific object class to very generic concepts that involve automatic feature learning, embedding of spatial context and object classification (e.g., approaches based on Artificial Neural Networks (ANN) and more specific deep learning concepts like Convolutional Neural Networks (CNN)).

The remainder of this book attempts to give an insight into the most important methodological components and concepts of these different approaches to automatic object extraction and scene analysis (Part I), while outlining the particular advantages and deficiencies of each concept. Based on this, selected approaches on object extraction, detection, classification and object tracking in remote sensing data are presented (Part II). Science develops rapidly, and none of these approaches will alone represent the current state-of-the-art. However, the approaches are selected in such a way that they capture the diversity of different attempts and views on the most important aspects of object extraction and scene analysis. The final part of the book (Part III) summarises and reviews these aspects and focuses on recent scientific trends and developments in the fields of computer vision and remote sensing, so that the reader will be able to obtain an insight into the basic facts, the range of opportunities, the potential challenges and also the limitations of this highly dynamic and complex field of computer and engineering science.

As the ultimate scope of this book is knowledge-based understanding of images, this book focuses more on semantic approaches to image interpretation rather than on learning-based "black-box" approaches. Especially under the light of the success of advanced deep learning schemes it seems necessary to put semantic concepts to image analysis back into focus more than ever. Because only by them, it will be possible to make the mechanisms involved in deep learning systems transparent and explainable (Roscher et al, 2019).

References

Gonzalez, R. and Woods, R. (2002). *Digital image processing* (2nd edition). Prentice Hall, Upper Saddle River, USA.

Hinz, S. (2004). Detection of vehicles and vehicle queues in high resolution aerial images. *Photogrammetrie Fernerkundung Geoinformation*, 3(4):201–214.

Hirschmüller, H. (2008). Stereo processing by semiglobal matching and mutual information. *IEEE Transactions on Pattern Analysis and Machine Intelligence*, 30(2):328–341.

Rosenfeld, A. and Kak, A. (1982). *Digital picture processing* (2nd edition). Academic Press, Orlando, USA.

Roscher, R., Bohn, B., Duarte, M. F. and Garcke, J. (2019). Explainable Machine Learning for Scientific Insights and Discoveries, available at: https://arxiv.org/abs/1905.088832019 (last access Aug. 25, 2019)

Notes

1. In the literature, the terms *scene analysis, image analysis, image understanding, computer vision, machine vision and robot vision* are often used synonymously.
2. Charged coupled device.
3. Complementary metal oxide semiconductor.
4. Here, we make no difference between digitally recorded images and scanned photographs.

Object, Data and Sensor Modelling

CHAPTER 2

Stefan Hinz

2.1 Sensors and Sensor Models

In the following, we sketch the characteristics of the most important remote sensing sensors on a very general and simplified basis in order to provide an overview of the basic components of sensor models. If needed for the understanding of particular image and scene analysis approaches, the application-oriented chapters of Part II of this book provide more details of specific sensors and sensor models.

2.1.1 Cameras and Image-based 3D Reconstruction

Apart from satellite remote sensing where so-called line scanners play a major role, frame cameras are the most important sensor for acquiring images from airborne, terrestrial or UAV[1] platforms. High-quality panchromatic or multi-spectral aerial images, captured during photogrammetric aerial flights, often serve as a primary data source. However, in recent years, rapid developments in the consumer camera market have allowed the design of much cheaper (multi-)camera systems that are still suitable for applications with high geometric and radiometric requirements.

Aerial or UAV images are often taken with high along-track overlap, ranging from 60% up to more than 90% overlap of consecutive images, as this model parameter depends primarily on the maximum frame rate and storage capacity onboard (Figure 2.1). Across-track overlap influences the flight duration and is thus more sensitive to the budget of a flight campaign than along-track overlap. Hence, across-track overlap typically ranges from 20% to 60% in photogrammetry, and occasionally more for specific applications. Due to automated flight navigation and, especially in photogrammetric campaigns, stabilised camera platforms, nearly the entire image bundle of a campaign can be captured in predefined viewing directions. While remote sensing and photogrammetric applications typically prefer nadir views, close-range and UAV images are often taken in oblique viewing directions, too. The primary sensor model parameters for determining the detailedness (i.e. the scale or geometric resolution) of an image are the exterior orientation parameters (position and attitude of the camera) and the interior orientation parameters of the camera, most prominently focal length, pixel size and lens characteristics such as radial/tangential distortion, spherical/chromatic aberration and vignetting. The procedure of camera

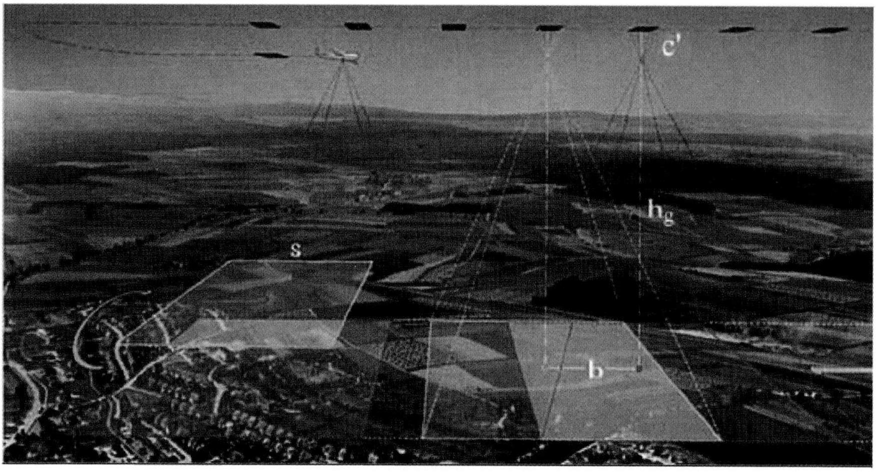

Figure 2.1 Tpyical set-up of a photogrammetric flight (from Albertz, 2009) with h_g being the hight above ground, b the basis, s the side length of an image projected on ground and c' the camera focal length.

calibration usually subsumes the determination and correction of these parameters (Hartley and Zisserman, 2008; Kraus, 2007; Steger et al., 2007), resulting in a geometric resolution of the calibrated images being able to capture object details in the range of a few centimetres or decimetres for UAV and aerial images and roughly 0.5 m to several metres (or even more) for satellite images.

The radiometric resolution of digital images is often 8 bit per channel, although the original radiometric resolution of the sensor is often higher. Reducing the radiometric resolution to 8 bit per channel is sometimes done, though, in order to apply more efficient image processing algorithms. A better radiometric resolution gives rise to a higher dynamic range of the recorded intensities, which eases the analysis of dark or shadowed areas in particular. During the development of photogrammetric digital cameras in the late 1990s, it was shown, that a higher dynamic range leads to significantly better accuracy of the resulting digital surface model (DSM)[2] in shadowed areas. Typical parameters further influencing the quality of the radiometric resolution are quantum efficiency, dark current, blooming, photon noise and charge transfer efficiency, which relate to the linearity and stochastic behaviour of the detector elements and the measuring units of the charge coupled device (CCD) and complementary metal oxide semiconductor (CMOS) sensors that transform the incoming electromagnetic radiation (EMR) into digital numbers. Luhmann (2010), for instance, gives further insight into this topic.

Most of today's cameras feature at least one panchromatic channel and three multi-spectral channels (red, green and blue), often with one or even more infrared channels covering the near and occasionally mid and far/thermal infrared bands. This so-called spectral resolution is expanded in the special case

and the unique identification of the reflection point, it is sufficient to record only one echo per emitted signal. On the other hand, cw laser systems require more energy to operate over long ranges so that they are less attractive for airborne platforms.

Conventional pulsed laserscanner systems for topographic mapping are based on time-of-flight ranging techniques to determine the range to the reflecting object. The time of flight is measured from the elapsed time between the emitted and backscattered laser pulses, with each pulse being only a few nanoseconds long. For objects such as vegetation showing more than one single scatterer inside the laser beam, more than one backscattered pulse may be detected per emitted pulse. Hence, for such objects, some sort of volume scattering can be reconstructed. Most airborne laserscanner (ALS) systems are able to capture, at a minimum, the range for the first and last detected backscattered pulses. Some systems acquire ranges for multiple backscattered pulses, up to the quasi continuous waveform of the echo per emitted pulse. Hence, these systems are also known as full-waveform laserscanners. In contrast to earlier large footprint systems like the spaceborne Geoscience Laser Altimeter System (GLAS) mounted on the Ice, Cloud and land Elevation Satellite (ICESat), small-footprint ALS systems illuminate only a small patch of the scene within the footprint (e.g. 20 cm diameter), yielding waveforms with distinct return pulses corresponding to specific surface characteristics.

This technique is better suited for long-distance measurements (more than 1 km), because less energy as for cw laser scanning systems is needed. At ranges of 300 m to 1 km, pulsed systems reach an accuracy of single points measuring 5–15 cm with a footprint diameter of about 20 cm. Pulsed laserscanners are as attractive remote-sensing measurement systems, because – depending on the system and platform – they can reach a high point density, thus mitigating the measurement noise by simple averaging, and, as stated above, they are able to record the full waveform of the received signal.

An example of a signal profile applied to multiple pulses is depicted in Figure 2.5. The waveform parameters for each detected pulse of this signal profile are estimated by a Gaussian decomposition method (the extracted features are given in the table), and the estimated waveform is shown below the original waveform. By comparing the range values in the table of Fig. 2.5, we see that the distance between the first and second pulse is about 10 m and between the third and fourth pulse about 2.5 m. The third pulse shows the highest amplitude, and the pulse widths of the first and second pulses are slightly lower than the pulse widths of the third and fourth pulses. It is not possible to classify the type of surface illuminated within a single beam corridor using the amplitude, pulse width and range properties alone. To assign each return pulse to a specific surface type, additional information is required, which can result from the 3D geometrical relationships of the returns within a point cloud.

Figure 2.6 visualises an example of a full-waveform data set of a test scene captured by scanning along the azimuth and elevation and recording the return

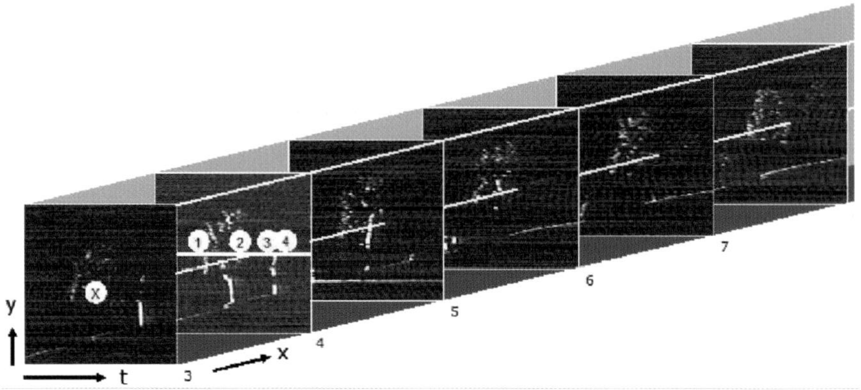

Figure 2.6 Vertical image slices with ground, vegetation and building structures.

intensity sampled over time t. When neglecting angular variations of the scan, the measured intensities as a function of time t sampled over the azimuth and elevation can be interpreted as a 3D data set forming a cuboid with Cartesian coordinates x, y and t. Sampling along the time axis can be recalculated into corresponding range values z. From this cuboid, various parameters of the echo pulse are achieved by combining echo properties and the spatial relationship of the returns. Figure 2.6 shows a set of image slices (yt planes). The second slice from the left ($x = 4$) shows vegetation in the centre (near range) and building structures on the right side (far range). Greyvalues correspond to the intensity of the signal. Note that, although this way of displaying the data suggests that a full 3D representation of the scene has been obtained, this is in fact not the case. Just like point clouds measured by conventional laserscanners, the data cube represents only 2.5D information. This is because of occlusion effects that are dependent on the target size in relation to the beam footprint size. It is possible that the laser pulse is mostly intercepted by and backscattered from the first illuminated surface along the propagation direction and that following surfaces along the laser vector are hidden, giving weak or no reflections. For instance, a tree with dense foliage may return only a single reflection response per laser pulse even though multiple surfaces are present after the first detected return along the path of the laser vector.

Much more detailed approaches towards exploiting the data and sensor characteristics of full-waveform laserscanners will be addressed in later chapters of this book.

2.1.3 Synthetic Aperture Radar

Like laser scanning systems, a synthetic aperture radar (SAR) system is an active sensor. With the success of the first environmental SAR satellites, ERS-1 and ERS-2, SAR technology has become very popular in the remote sensing community. In particular, because SAR is an active system and largely independent of weather conditions, it is an attractive imaging technology for rapidly acquiring area-wide information for remote regions prone to harsh weather conditions. The first SAR satellites were only able to map areas with a geometric resolution of 20–30 m, but, since 2007, a new class of space-borne SAR sensors (e.g. TerraSAR-X, TanDEM-X, SAR-Lupe or Cosmo-SkyMed) provide images with 1–3 m spatial resolution (or even below) in special spotlight modes. One major focus of these missions is the improvement of mapping of rural and urban areas. The development of methods to automatically derive detailed cartographic information has undergone much development in recent years. Furthermore, due to the fact that high-precision repeat-pass tracks and the TanDEM-X mission provide high-resolution interferometric SAR data, methods for automatic 3D information extraction have come into play. These will be addressed in Chapter 11 of this book.

In the following we give a short description of the image formation process of SAR, as it involves the most important sensor characteristics mandatory to consider for further image analysis tasks.

The radar emits frequency-modulated signals – so-called chirps – with a predefined pulse repetition rate (PRF) in a side-looking, oblique imaging geometry, and records the echoes scattered at the objects on the ground; see Figure 2.7 (upper left) for an illustration of the radar imaging geometry. The echoed chirps are correlated with the sent (reference) chirp, eventually yielding a compressed pulse-shaped signal whose width is mainly determined by the chirp's bandwidth (Figure 2.7, upper right). The travelling time of the signals is proportional to the distance to the objects and defines the image dimension perpendicular to the flight direction, the so-called range or across-track coordinates. The second dimension, the azimuth or along-track, is simply aligned with the flight direction.

While the resolution in the range direction is determined by the chirp band width and is typically in the (sub-)metre area, the resolution in the azimuth direction of the raw data depends on the antenna's real aperture characteristics (antenna length, carrier wavelength and range) and is impractically coarse for geospatial applications (Fig. 2.7, lower left). Hence, to enhance the azimuth resolution, the well-known SAR principle is applied; that is, the motion of the real antenna is used to construct a very long synthetic antenna by exploiting each point scatterer's range history recorded during a point's entire observation period (Fig. 2.7, lower right), also forming a chirp-like signal (i.e. the so-called azimuth chirp). An optimally focused image is thus obtained by complex-valued correlation of the reference azimuth chirp and the recorded azimuth chirp. To construct the reference azimuth chirp correctly, the actual range or phase history of each target in the image must be known, which can be inferred from sensor and scatterer position. Usually, the time dependence of the scatterer position is ignored. From an algorithmic point of view, this concept of azimuth focusing of the SAR image is performed using sophisticated matched

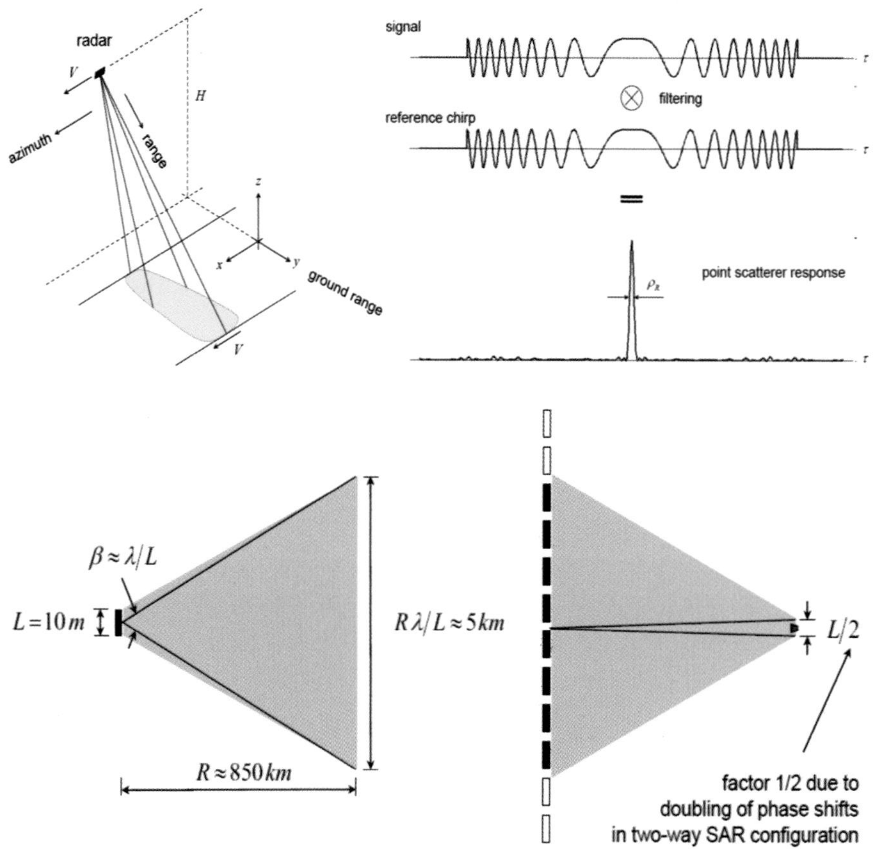

Figure 2.7 Radar imaging geometry (upper left) and compression of the chirp signal sent in line of sight direction (also called slant range) into a pulse (upper right). Forming a high resolution synthetic apertur in azimuth direction based on capturing low resolution echos sent from the moving real aperture (lower left) and processing them in such a way as if one long antenna had been used (lower right).

filter concepts (Bamler and Schättler, 1993; Cumming and Wong, 2005), also commonly referred to as the stationary-world matched filter (SWMF) concept.

Since the length of the synthetic aperture increases proportionally with flying height, the resolution in the azimuth direction purely depends on the length of the physical antenna given a sufficiently large PRF to avoid aliasing. Figure 2.7 compares a real aperture and formation of a synthetic aperture for a typical spaceborne SAR configuration, and Figure 2.8 shows examples of focused SAR images.

The image formation process of SAR involves both the amplitudes and phases of the recorded signal, so SAR belongs to the category of coherent imaging techniques (like, e.g., tomography). Amplitude variations of SAR images thus follow

Figure 2.8 High-resolution SAR scene with 1 m geometric resolution (top) and mountainous area of medium-resolution SAR (SRTM, bottom).

a Rayleigh distribution, and intensity values (i.e. the square of amplitude) follow an exponential distribution, leading to the so-called fully developed (multiplicative) speckle noise (Bamler and Schättler, 1993; Cumming and Wong, 2005). Yet these relations hold only under the condition of an almost infinitely high number of scatterers in each resolution cell, which can be assumed for medium- and low-resolution SAR but is rarely fulfilled for high-resolution SAR images.

It is important to note, that SAR is a 2D imaging sensor, although distances to objects are measured. Yet, in contrast to laserscanners, it is not possible to measure the pointing direction – here called elevation angle – of the radar beam with sufficient resolution. Only by using multiple SAR images can the elevation angle be determined with reasonable resolution. This leads to the typical SAR characteristics that all scattered signals with the same distance to the sensor are superimposed in the same resolution cell, also called "layover", leading to the impression that steep or vertical objects lean towards the sensor (Figure 2.8, bottom). The opposite effect, the radar shadow, is comparable to occlusions in optical images and appears as scatterless resolution cells.

The above description of sensors and sensor models used in remote sensing only touches on the most important measuring techniques and their typical characteristics. In fact, many more remote sensing technologies exist to capture

2D, 3D or *n*D information about objects. Further popular sensors include structured light sensors (such as Kinect) or time-of-flight video cameras. Depending on the requirements of certain image analysis tasks, some issues concerning these sensors will be tackled in later chapters.

Besides the sensor models, the complexity of the recorded scenes and of the data to be treated is of essential importance when designing an image analysis system. To enable successful and efficient object extraction, the model and strategy must be adapted to this complexity. In the following, we will outline these principles.

2.2 Data Content and Complexity

As far as images are concerned, the term *complexity* is closely related to the visual impression of the individual viewer. Figure 2.9 shows some example subsets of aerial images. Viewers naturally feel that the subsets of rural areas can be more easily interpreted and are therefore less complex than those of urban areas. The subjective impression of complexity makes its quantification – ideally with a single measure – extremely difficult (Shufelt, 1999). However, in order to allow at least a qualitative categorisation, a number of factors can be mentioned that contribute to the complexity to a greater or lesser extent. First of all, the content of the scene in the real world and the observability of the content, defined by the sensor, are to be distinguished (Mayer, 1998).

Important factors that contribute to the complex content of a scene from the real world (based on unit area) are as follows:

- *Number of objects and object classes*: The more individual objects from different classes (such as buildings, roads and water bodies) that are present in a scene, the more comprehensive and variable is the scene.
- *Intra-class variability* (i.e. the variability within an object particular class): This is basically defined by largely varying characteristics of objects belonging to the same class. The class "vegetation", for example, has a much higher intra-class variability than the class "road marking".
- *Distinctiveness of object classes*: Different object classes may be characterised by pronounced differences in their characteristics (in terms of different means, standard deviations, different texture features, etc.). However, distributions between object classes may also be similar (e.g. roads and flat roofs made of concrete), thus reducing inter-class variability and complicating their separation.
- *Relations between objects*: The more different relations between objects that occur, the more variable is the appearance of the objects.

Besides these, complexity is also affected by sensor-related factors. In the following, specific characteristics are summarised (effects due to long time intervals between image acquisitions are neglected):

Figure 2.9 Complexity of images depending on their resolution. Rows (left to right): increasing complexity of the content of the scene. Columns (top to bottom): decreasing observability (resolution: high, medium, low).

- *Imaging geometry*: This determines which part of the scene is recorded under which kind of geometric projection. Aerial frame cameras produce images with central perspective projection in all viewing directions. However, the images of CCD line sensors are characterised by a central perspective projection only perpendicular to the flight direction, and show a parallel projection along the flight direction.
- *Image scale/ground sampling distance (GSD)*: The smaller the GSD, the smaller the objects and details that are observable and the higher the complexity. For images of man-made objects, a GSD larger than 1 m is commonly defined as low, between 0.25 m and 1 m as medium, and less than 0.25 m as high-resolution data (Mayer, 1998). The image scale is given by the ratio of the focal length of the camera and the flying height, and it may differ from pixel to pixel depending on the viewing direction and terrain relief.
- *Spectral resolution*: The more channels are available, the more information about the spectral properties of the objects is included in the image. One typically divides images into pan-chromatic images (one band), multi-spectral images (roughly up to a dozen of bands) and hyperspectral images (up to several hundreds of bands).

- *Radiometric resolution*: This specifies the dynamics of a sensor (i.e. how sensitively the energy of the EMR can be recorded). While 8 bit (256 grey-values) per channel is typically the usual quantisation, the radiometric resolution of many newer sensors in remote sensing can already capture EMR dynamics with 12 bit and more.
- *Temporal resolution*: Remote sensing data sets are acquired with different repetition cycles. Although short videos can be also taken from space meanwhile, a second image of the same point on Earth may be available only after a couple of hours or days after the first, or may be acquired years after the first[3]. In both cases, temporal resolution may complicate the analysis, either because the *availability* of numerous images inflates the data volume, or because the *lack* of images may lead to there being insufficient information (e.g. summer conditions in vegetation analyses). On the other side, if the temporal behaviour of an object class can be modelled – such as the phenomenology of vegetation – multi-temporal data may significantly facilitate object extraction.

Primarily, a higher complexity of data complicates the process of image analysis. The variety and variability of different objects and object classes, as well as the relations among them, can easily lead to missing extractions or misinterpretations. A high degree of complexity, however, is not necessarily an obstacle to object extraction. Often it is the details of an object and its relations to other objects that make it identifiable. A typical classroom example for this is shown in Figure 2.10. The sub-object "chimney", for example, gives a clear indication of a house roof or a building, and the relation "vehicle driving on a road" implies a hidden lane and thus a road. These examples prove that a powerful image analysis system is characterised by an appropriate adaption of model and strategy to the complexity of the scene and data.

2.3 Object Modelling

2.3.1 Introduction

Any object detection is based on a model of the object to be extracted. As stated above, only by means of a model can the implicitly contained information of an image or other remote sensing data be related to the semantics of the corresponding subset of the scene in the real world. Therefore, the aim of modelling must be to structure, formalise and represent the available knowledge about the object in order to successfully reconstruct the semantics and the properties of an object on the basis of the raw data.

In the remainder of this book we do not make a sharp distinction between objects and object classes. A certain object such as a specific car in an image is always understood as a typical representative of the object class "car". If truely a specific instance of an object class is meant, we indicate this accordingly.

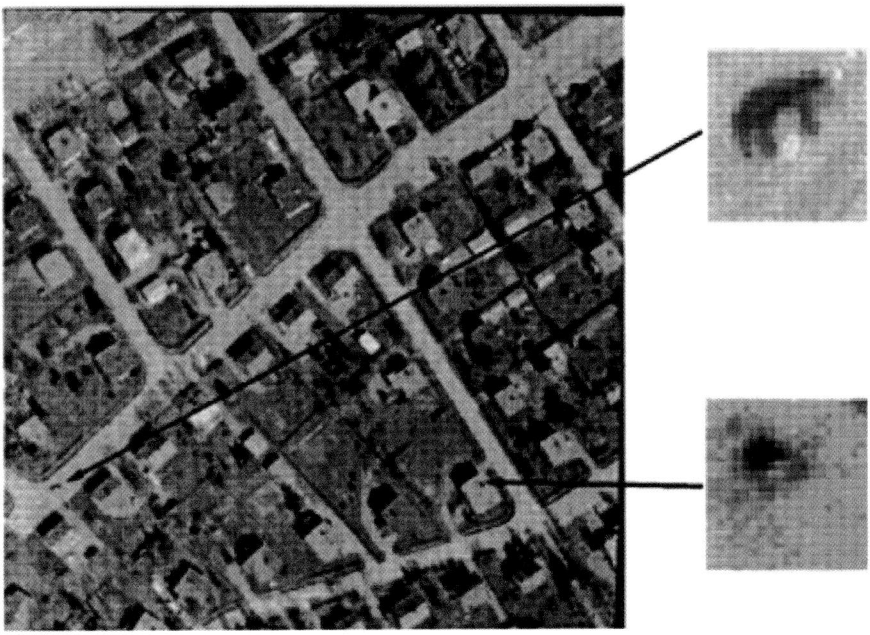

Figure 2.10 The context information of the left image allows the identification of objects in the two image patches on the right.

There are different approaches to the components of modelling, which will be described in more detail in the following subsections. Even if not every component of a model is used for automatically detecting and characterising objects in the data or even if a model is not fixed *a priori* but learned from sample data, it is nonetheless helpful to (conceptually) set-up an object model on the level of the real world and, by incorporating a sensor model, to transfer this into an image model of the object (or more generally, a data model of the object). Doing so has the advantage that the appearance of the object and its properties in the image can be explicitly connected to symbolic descriptions and hierarchical structuring of objects – an approach that supports the understanding of the performance and deficiencies of the implemented image analysis algorithms. Moreover, especially when dealing with complex scenes, it is necessary to integrate knowledge about the surrounding influences on the object and its appearance in the model. This is realised through the context model.

The aim of modelling is to provide a description of an object, which is structured, preferably consistent and formally representable in the computer. One usually distinguishes an *explicit* and *implicit* representation of the model knowledge. In the following sections, we describe the basic principles of these modelling approaches starting with explicit modelling.

2.3.2 Explicit Modelling

An explicit description of the model knowledge about objects and their relations is advantageous because the results generated by the algorithm can be analysed more easily, and can therefore be directly used for predicting the system's behaviour for other data sets. It is, however, limited to those cases for which the particular object classes, data and sensor characteristics can be modelled properly.

The knowledge contained in the model can be structured, based on different aspects that complement one another, as discussed in the following.

2.3.2.1 Structuring in Levels

A first categorisation is often done in the form of different levels. As has been outlined before, structuring model knowledge can be achieved by dividing the model into an object model, which corresponds to a description using concepts of the real world, and an image model, which describes the appearance of the object in image space. Transfer of the object model into image space is done by the sensor model, which captures the geometric and physical transformations during the sensing process. The two main levels are thus the "real-world" level and the "image" or "data" level. At each level an object can be hierarchically subdivided into sub-objects by specialisation and part-of relations based on specific criteria. At the image level, these criteria are used to relate image features that can be extracted from the data to the sub-objects; that is, this level is closely related to the capabilities and efficiency of image-processing algorithms. In contrast, the *function* of the objects plays an important role in the "real world" level. For example, existing highways must have multiple lanes to allow high-speed and safe overtaking, and must correspond in curvature and slope to given construction guidelines. Such functional properties can be very favourably used for structuring the object into its parts and for setting plausible parameter values for geometrical and physical object properties.

2.3.2.2 Structuring According to Semantic Abstraction and Resolution

A model can be divided into different levels of resolution. In the image space, this means that the image model is available at different scales (i.e. in discrete realisations of the continuous scale–space). The most common scale-space is the linear or Gaussian scale-space, where lower-resolution levels of an nD original signal are produced by convolution with the Gaussian smoothing operator:

$$g(x, t) = \frac{1}{(2\pi t)^{n/2}} \cdot e^{\frac{-x^\top x}{2t}} \tag{2.1}$$

in which x is 2D ($n = 2$) for conventional images and the smoothing or scale-space parameter t represents variance σ^2. After smoothing, resampling according to the Nyquist theorem is usually carried out, so the lower resolution of the images is expressed by a larger GSD.

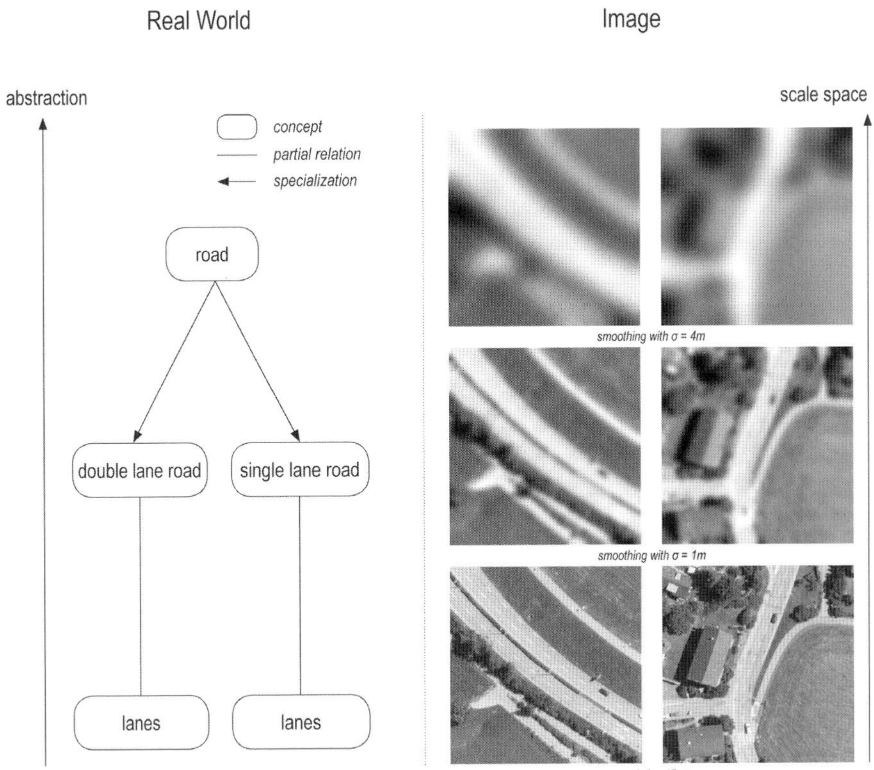

Figure 2.11 Levels of abstraction formulated as a semantic network (left) and the corresponding scale-space (right). With increasing smoothing σ of the image, details such as markings and cross-walks, and later larger structures such as vehicles and the constructional separation of the lanes, are eliminated. The results of these events can be assigned to different levels of abstraction in the real world.

As shown in Figure 2.11, a scale-dependent modelling at the level of the real world corresponds to a subdivision of the object model into several levels of abstraction. This kind of structuring emphasises the generalisation of knowledge at higher levels of abstraction and underlines the different characteristics of the object at different scales. The advantages of this kind of modelling were particularly pointed out in classical works on semantic object extraction such as Mayer (1998) and Mayer and Steger (1998).

Generally speaking, this indicates that an extraction solely using images of the best resolution does not necessarily lead to the best results. Each object property has a certain scale range in which it is particularly observable. Thus scale-based modelling not only contributes to a better structuring and efficient use of knowledge, but is also an important basis for successful extraction, especially when the results from different scales are combined into a scale-dependent extraction framework.

The combination of certain abstraction levels with their corresponding scales in the image domain is usually done in advance by fixing a certain discrete number of scales. Some research activities are also directed towards an automated analysis of so-called scale-space events, be it numerically or analytically (Mayer and Steger, 1998). In the case of complex topographic objects, automatic scale estimation, which is necessary in this context, only succeeds exemplarily. However, for the extraction of simple image features such as lines and edges it can be formulated more generally (Lindeberg, 1998).

2.3.2.3 Structuring by Context

In natural environments, an object is often influenced by certain neighbouring objects or its entire surroundings. It is therefore not sufficient to model the object as an isolated unit. Roads, for example, often appear in aerial images as a bright grey surface. In the case of shadows cast by adjacent buildings or trees, the bright surface becomes dark and therefore it no longer corresponds to the assumptions of the model. Knowledge about such influences can be formulated by a so-called context model. That means that the object model is extended by relations to other objects, or, in the case of a more comprehensive modelling, it is embedded in a model for the complete scene. Taking into account the original concept of Strat (1992), context in the concept of explicit modelling describes knowledge about the environment of the object to be extracted and the resulting impact on its shape and appearance. The interaction of object and environment is particularly evident when due to impacts from the immediate neighborhood. However, because dependencies on the environment occur also when taking into account large-scale object properties and object relations, analogous to Mayer (1998) and Baumgartner (2003) a subdivision into *local and global context* is made. For detailed context modelling it must be pointed out that interactions of object relations could also appear. For example, a car passing through the shadow of a building practically does not cast an additional shadow on the road. The shadow relation between the road and the vehicle is thus erased by the shadow relation between the road and the building.

The objective for model global context knowledge is related to the observation that, depending on the particular large-scale environment, specific properties and local contexts of the model are particularly striking and others become less important. In open and rural areas, for example, there is, from an economic point of view, usually only one optimal connection between two points on the road network, while in urban areas, due to dense building development, several nearly equivalent alternatives often exist (van Cleynenbreugel et al., 1990). Model properties, arising from the connecting function of roads, therefore have a much higher relevance in open and rural areas than in urban environments. The same applies to the importance of various local contexts. By including such global context areas – often "settlement", "open landscape" and "forest" – into the model, certain components of a complex model can be extracted more effectively and certain model parameters can be adapted to the particular context area (van Cleynenbreugel et al., 1990). This leads to a self-adaptive extraction system with a specially adapted model and specially developed strategy for each global context area, so that even very large scenes can be handled by effective models and efficient strategies. Summarising, local context describes the interactions of an object with objects in the immediate neighborhood. In contrast, the global context describes the large-scale specification of object classes, and therefore enables focusing on the typical characteristics, appearances and local contexts within the particular global context. This kind of context modelling profoundly supports the flexibility of the extraction strategy as, depending on different local and global contexts, the extraction strategy can provide several alternatives and thus allows dynamic control (regulation) of the extraction process.

2.3.2.4 Representation and Formalisation of Model Knowledge

An essential point for explicit object modelling in its classical sense is the representation and formalisation of the knowledge contained in the model. The form of knowledge representation should be chosen in order to fairly represent the model knowledge on the computer without any restrictions in terms of size and structure. At the same time, however, it should ensure a high degree of clarity for a detailed examination and testing of the image analysis system. In Brachman (1979) five levels of knowledge are presented: linguistic, conceptual, epistemological, logic and implementation levels. The linguistic level is based on natural language representation and is considered inappropriate for object extraction due to the absence of a clear formalisation of knowledge representation. The logic and implementation levels, however, are subject to very strong constraints. They can easily lead to complex and sometimes contradictory (intermediate) interpretations in a more comprehensive image analysis system, as shown, for example in McKeown et al. (1989). The epistemological level allows an application-independent representation of knowledge and therefore is suitable for a comparative analysis of different approaches, as in Mayer (1998), while the conceptual level – relating to object extraction – has the advantage that it allows an object-centred and thus a problem-oriented representation, but at the same time enforces the need for a clear formalisation of the knowledge representation. The conceptual level is often used for explicit modelling in early and pioneering object extraction systems; the specific representation formalism, which is particularly suitable and frequently incorporated in these systems, is a so-called semantic network (Brachman, 1977, 1979; Liedke et al., 2001; Sagerer, 1982; Woods, 1975). In addition, semantic networks show the advantage that they can be easily formulated as Bayesian networks – or more generally as so-called graphical models – so the relations between parts of objects and their appearances in the image can be attached with conditional probabilities among themselves. In this way, unavoidable uncertainties during the extraction process (in assigning image features to specific object classes) can be modelled and handled adequately. Besides classical Bayesian theory, a variety of mathematical and stochastic frameworks could be applied, such as Dempster–Shafer, evidence or fuzzy theory (Shafer and Pearl, 1990; Zadeh, 1989). However, if using these theories within the conceptual framework of an extraction system, it is particularly important to ensure that both modelling and extraction strategy do not violate the underlying assumptions of these theories.

2.3.3 Implicit Modelling

For many applications it is sufficient, easier or even better not to perform an explicit transition from the object model into the image space, but to only utilise an image model. For such approaches, implicit or appearance-based modelling comes into play. Like in the case of explicit modelling, the utilisation of different resolution and context plays an important role. The main difference from the concepts described above concerns the direct use of image patches and derived features and parameters therefrom as a model of the objects and relations.

The easiest case of implicit modelling uses a template image of the object and matches this template to the search image with any kind of similarity metric (e.g. greyvalue differences, cross-correlation, etc.). These techniques assume that one single instance of an object represents the whole object class properly. Partial occlusions, small deformations, clutter, illumination differences between template and search image, and so on are supposed to be accommodated by the application of sophisticated matching techniques being invariant against these influences (e.g. rotation and scale-invariant matching, feature-based matching, implicit shape matching, deformable matching, etc.).

While such matching techniques test the similarity of the template and search image by applying one or more thresholds of the technique's specific metric(s), *stochastic appearance-based approaches* rely on a statistical model of the appearance of an object (and occasionally the object's relations). As in any kind of supervised classification scheme, the model is learned from extracted features of sample data and represented in a multi-dimensional feature space. An unknown image patch is eventually tested against the model by evaluating some sort of distance – in feature space – between the model and the features extracted from the unknown patch. The model may be parameterised in feature space with analytical probability density functions (e.g. for Gaussian normal distributions) or be represented by discrete data as in the case of support vector machines (SVMs). A further distinction between different classification techniques can be made on the type of modelling. *Generative* classification approaches learn models of object classes and test whether an unknown image patch belongs to a certain class (based on a given distance metric), whereas *discriminative* classification approaches model only boundaries between different classes, thus not taking the shape or distribution of an object class in feature space into account, and test on which side of the boundary the features of an unknown image patch eventually lie. The following chapters in Part I of this book give a more detailed introduction into feature extraction and classification techniques.

Even more sophisticated appearance-based approaches include the stochastic modelling of local contexts (e.g. Markov random fields (MRFs) and conditional random fields (CRFs)), the distribution of relative distances and orientations between objects (e.g. component-based matching) and the learning, selection and fusion of the best suitable features (e.g. CNNs). It needs to be noted, however, that a higher complexity of the model to be learned usually calls for significantly more training data to learn the model and the involved coefficients and parameters to be sufficiently precise. Thus, a trade-off between the availability of adequate sample data and the mandatory model complexity must be observed in any case when designing an appearance-based extraction system. Later chapters of this book give an insight into this issue.

2.4 Extraction Strategy

In the context of object extraction, the extraction strategy is defined as "a plan for the recognition of objects in images by matching data to a model using various methods" (Mayer, 1998). This statement reveals that the strategy consists of two basic, interconnected elements: on the one hand, the development of the extraction process, and on the other hand, the choice of the particular image and data processing methods for its realisation. Mostly, the choice of methods highly depends on the application, so the methods relevant for a particular object extraction system will be described in the application-oriented chapters. In particular, the planning of the extraction process controls at what stage which kind of knowledge about the object is used for extraction and how the results of intermediate extractions are integrated in the further procedure. In this context, one usually distinguishes data-driven, model-driven and hybrid/mixed methods. In Figure 2.12, the basic steps of data-driven and model-driven approaches are compared. It has to be noted that this is an idealised scheme, as is the following description. For practical applications, only the use of the hybrid approach is realistic. Differences exist, however, in the weighting of data-driven and model-driven parts.

2.4.1 Data-driven Approach

The data-driven approach starts with purely iconic pre-processing steps, where the input and output of processing are of the same data type (image). In the next step, features from the images are extracted and combined to form superior structures (cf. Chapter 3). This process is continued in 2D and 3D until structures have been built that sufficiently meet the requirements of the model, and thus enable an unambiguous assignment to the model. To apply this procedure successfully it is mandatory that, at each step, new information is gathered,

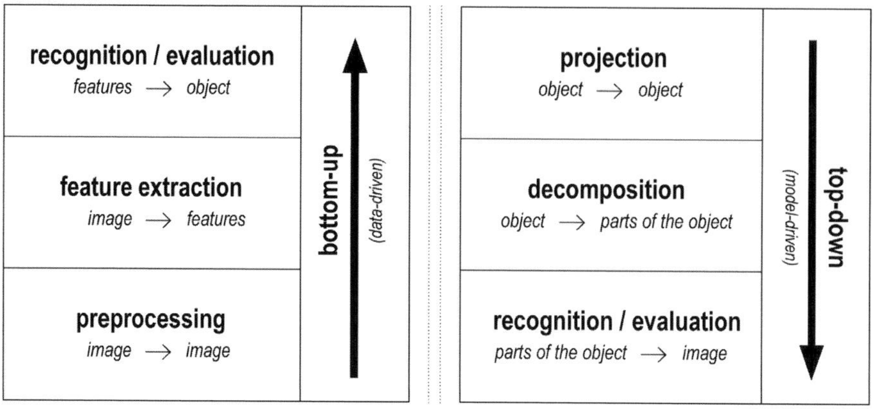

Figure 2.12 Procedures of image analysis in an idealised form: data-driven (left) and model-driven (right).

thereby constructing the complete object successively from the extracted parts. An advantage of this procedure is that a generic object model is sufficient; that is, there is no need for a pre-defined parametric model because the model components are directly obtained from the data by feature extraction and feature grouping methods.

2.4.2 Model-driven Approach

Within a purely model-driven approach, a (3D) object model is projected into the image or data domain via the sensor model, basically at every location and for each particular view onto the object. At each projection step, a hypothesis for the image-like appearance of the object is generated. The fitting of a hypothesis for a particular view and a particular location in the image is evaluated by specialised similarity metrics – usually after decomposing the object into individual object parts. A necessary precondition for this procedure is that the model and its appearance in the image can be described very specifically, so that a hypothesis in the image space corresponds to the actual appearance of the object. The advantage of the specific description is that, on the one hand, the search space for assignments can be strongly limited, especially when external knowledge about the scene is already available (e.g. a geospatial database) and, on the other hand, effects like occlusions, small distortions, background clutter or partially lacking data have less impact compared to the data-driven approach.

2.4.3 Hybrid Approach

A hybrid processing combines data-driven and model-driven aspects. In the case of interpretation of complex remote sensing data this is the only feasible approach. Decisive for the efficiency of a hybrid extraction system is the choice

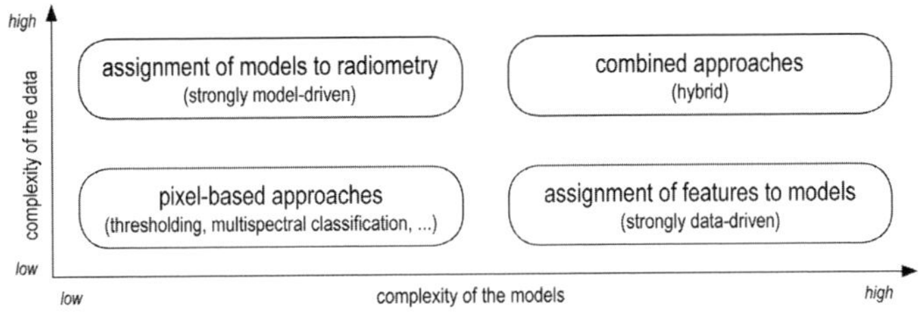

Figure 2.13 Efficient methods for different levels of complexity of data and models (according to Suetens et al., 1992; Mayer, 1998).

of "turning points" – when to move from data-driven processing to model-driven processing and vice versa. A typical approach is to start with data-driven feature extraction in order to obtain reliable hypotheses for object parts that can be easily and unambiguously extracted. These are evaluated for plausibility in a model-driven manner and used to generate new hypotheses, which are then verified or falsified by assigning newly extracted features. For example, the boundary polygon of a flat roof is relatively easy to extract from aerial images. Reliable roof hypotheses can thus be generated by (more or less) purely data-driven processing. Then, with the help of roof hypotheses, the search area of other object parts such as facades and shadow edges can be predicted and restricted by model-driven hypotheses generation. Hence, on the one hand, extracted object parts verify the roof hypothesis; on the other hand, they also allow a more detailed reconstruction of the buildings (Kim et al., 2001). A major advantage of the hybrid processing lies in the fact that additional knowledge can be implemented by the appropriate choice of turning points, that is, the (engineering type of) knowledge about typical capabilities and limitations of image processing methods concerning a certain task.

2.4.4 Suitability of the Extraction Strategy

The suitability of an extraction strategy can be categorised depending on the complexity of the model and the used data, as follows (Figure 2.13). For simple data and simple models, segmentation or classification approaches suffice for the extraction of objects. This can be seen in the widely used and successfully applied pixel-based multi-spectral classification methods, for instance, to classify satellite images of rural scenes. The assignment of models to radiometry – thus a strong model-driven strategy – can be applied if the data are characterised by high complexity, whereas the model itself can be described easily and specifically or the search space can be strongly limited. This aspect plays an essential role for the extraction of objects that can be easily parameterised in a CAD-like manner, such as vehicles, for instance. In the case of lower data complexity but higher model complexity, a strategy assigning geometric and radiometric features to the model is suitable. As the content of the underlying data is quite simple, feature extraction and segmentation can provide clear evidence for an object. Thus, the extracted features can be successfully assigned even to complex models. If both data and model are complex, a hybrid strategy for object extraction should be utilised that is adapted to the particular characteristics of the data and the model (Suetens et al., 1992).

Various methods are used for the realisation of a strategy, including, on the one hand, relatively application-independent algorithms of digital image processing such as radiometric, geometric or morphological transformations, segmentation, feature extraction, polygonal approximation, etc., and on the other hand highly application-dependent methods for (multi-sensorial) data fusion and grouping of substructures. Generic search algorithms (Geman and Geman, 1984; Vosselman, 1992) and assessment schemes (Shafer and Pearl, 1990; Zadeh, 1989), which ostensibly do not depend on the application, represent another group of methods.

2.5 Conclusions

As has been pointed out in the Introduction, remote sensing image analysis is both a vivid and complex field of research. Despite significant advances – ranging from the early design of "semantic networks" to the recently pushed deep-learning methods like CNNs – it has become clear that there is no golden standard way to design an object extraction system. This chapter has aimed to provide a broad view of this issue to systemise the tasks involved and to structure this complex problem into smaller tasks and challenges that can be handled more easily.

The chapter started by outlining the principal objectives of image analysis. Although certainly a daunting challenge – perhaps never fully achievable – it is necessary to attempt to make a computer "see" what a human operator sees. However, this aim is challenged by image complexity arising from the properties of the imaged world and the properties of the sensor. Three sources of information – radiometric, geometric and semantic – help human recognition to grasp this complexity and this has to be translated into the realm of computers. This is a main aim in designing an image analysis framework and is based on two steps. First, the analyst needs to define an object model according to levels, resolution and context, and represent both model knowledge and the uncertainties associated with it within the framework. Second, an extraction strategy has to be elicited. Here, taking account of the complexity of the data and the complexity of the object model, the analyst can choose from data-driven, model-driven or hybrid approaches.

The application-oriented chapters within this issue will outline how deliberate decisions on these crucial questions can render remote sensing image analysis a useful tool to gain insights into a wide range of analyses in both human and natural environments.

References

Albertz, J. (2009). *Einführung in die Fernerkundung* (4. Auflage). WBG (Wissenschaftliche Buchgesellschaft), Darmstadt, Germany.

Bamler, R. and Schättler, B. (1993). *SAR geocoding*. Wichmann, Karlsruhe, Germany.

Baumgartner, A. (2003). *Automatische Extraktion von Strassen aus digitalen Luftbildern*. Dissertation, German Geodetic Commission (DGK), Vol. C, No. 564, Munich, Germany.

Brachman, R. (1977). Whats in a concept: Foundations for semantic networks. *International Journal of Man–Machine Studies*, 9:127–152.

Brachman, R. (1979). *On the epistomological status of semantic networks*. Academic Press, New York, USA.

Cumming, I. and Wong, F. (2005). *Digital processing of synthetic aperture radar data*. Artech House, Boston, USA.

Geman, S. and Geman, D. (1984). Stochastic relaxation, Gibbs distributions, and the Bayesian restoration of images. *IEEE Transactions on Pattern Analysis and Machine Intelligence*, 6(6):721–741.

Hartley, R. and Zisserman, A. (2008). *Multiple view geometry in computer vision*. Cambridge University Press, Cambridge, UK.

Hirschmüller, H. (2008). Stereo processing by semiglobal matching and mutual information. *IEEE Transactions on Pattern Analysis and Machine Intelligence*, 30(2):328–341.

Kim, Z., Huertas, A., and Nevatia, R. (2001). A model-based approach for multi-view complex building description. In Baltsavias, E., Grün, A., and van Gool, L., editors, *Automatic extraction of man-made objects from aerial and space images (III)*, pages 181–193. Balkema Publishers, Lisse, The Netherlands.

Kraus, K. (2007). *Photogrammetry: Geometry from images and laser scans*. de Gruyter, Berlin, Germany.

Liedke, C.-E., Bückner, J., Pahl, M., and Stahlhut, O. (2001). Knowledge based system for the interpretation of complex scenes. In Baltsavias, E., Grün, A., and van Gool, L., editors, *Automatic Extraction of Man-Made Objects from Aerial and Space Images (III)*, pages 3–12. Balkema Publishers, Lisse, The Netherlands.

Lindeberg, T. (1998). Edge detection and ridge detection with automatic scale selection. *International Journal of Computer Vision*, 30(2):117–156.

Luhmann, T. (2010). *Nahbereichsphotogrammetrie: Grundlagen, Methoden und Anwendungen*. Wichmann, Berlin, Germany.

Mayer, H. (1998). *Automatische Objektextraktion aus digitalen Luftbildern*. Habilitation, German Geodetic Commission (DGK), Vol. C, No. 494, Munich, Germany.

Mayer, H. and Steger, C. (1998). Scale-space events and their link to abstraction for road extraction. *ISPRS Journal of Photogrammetry and Remote Sensing*, 53(2):62–75.

McKeown, D., Harvey, W., and Wixson, L. (1989). Automating knowledge acquisition for aerial image interpretation. *Computer Vision, Graphics, and Image Processing*, 46:37–81.

Sagerer, G. (1982). *Darstellung und Nutzung von Expertenwissen für ein Bildanalysesystem*. Springer, Berlin, Germany.

Shafer, G. and Pearl, J. (1990). *Readings in Uncertain Reasoning*. Morgan Kaufmann Publishers, San Francisco, USA.

Shufelt, J. (1999). *Geometric constraints for object detection and delineation*. Kluwer Academic Publishers, Boston, USA.

Steger, C., Ulrich, M., and Wiedemann, C. (2007). *Machine vision algorithms and applications*. Wiley-VCH, Weinheim, Germany.

Strat, T. (1992). *Natural object recognition*. Springer, New York, USA.

Suetens, P., Fua, P., and Hanson, A. (1992). Computational strategies for object recognition. *ACM Computing Surveys*, 24(1):5–60.

van Cleynenbreugel, J., Fierens, F., Suetens, P., and Oosterlinck, A. (1990). Delineating road structures on satellite imagery by a GIS-guided technique. *Photogrammetric Engineering and Remote Sensing*, 56(6):893–898.

Vosselman, G. (1992). *Relational matching*. Springer, Berlin, Germany.

Woods, W. (1975). What's in a link: Foundations for semantic networks. In Bobrow, D. G. and Collins, A., editors, *Representation and understanding: Studies in cognitive science*, pages 35–85. Academic Press, New York, USA.

Zadeh, L. A. (1989). Knowledge representation in fuzzy logic. *IEEE Transactions on Knowledge and Data Engineering*, 1(1):89–100.

Notes

1. UAV indicates unmanned aerial vehicle, sometimes also called an unmanned aerial system (UAS) or drone.
2. We distinguish DSM (digital surface model) and DGM (digital ground model). The DSM includes the complete surface of the imaged scene, while the DGM reflects the ground with houses and vegetation/trees removed. Other terms like DEM (digital elevation model) or DHM (digital height model) subsume both DSM and DGM, and are used where the distinction is not necessary.
3. Of course, the frequency with which images are acquired is *not* a property of the images *sensu stricto*. However, temporal availability does create complexity and should be considered in image analysis frameworks.

CHAPTER 3

Feature Extraction from Images and Point Clouds: Fundamentals, Advances and Trends

Martin Weinmann

A large variety of applications currently focus on the exploitation of 2D imagery or 3D point clouds as input data. In order to obtain such data, numerous sensors may be involved, and due to technological advancements their capabilities have steadily improved. Although the availability of the different kinds of sensors, such as digital cameras, thermal cameras or 3D sensors, offers an appropriate digitisation of real-world objects or scenes, the direct use of the captured data often remains impractical for applications. Since the large amount of available data typically increases the computational burden with respect to both processing time and memory consumption, some applications may significantly benefit from using only as much information as required for solving the particular tasks, instead of all the available information. On the other hand, some applications could benefit from transforming the captured data into a different representation that directly indicates specific properties of an object or specific properties of the scene. Consequently, finding an appropriate representation of the captured data is desirable, and the extraction of adequate features has become a basic requirement not only for extracting, describing and comparing objects, but also for tasks such as image or point cloud registration, data retrieval, scene analysis and the autonomous navigation of aerial or ground vehicles.

In order to become aware of the required features, we should consider a number of fundamental issues: What makes an object memorable? According to which criteria could we recognise the same or similar objects? How could similarity be described? The visual perception of an object depends on its size and shape, which can also indicate a robust or fragile structure. This impression is strengthened by the colour of the object as well as the reflectance behaviour on the object's surface, both of which often provide a strong hint regarding the material and value of the object. Furthermore, the homogeneity of the object surface and thus its textural properties may serve to distinguish between different materials. A feature thus describes a special (local or global) property by which real or abstract elements or objects can be distinguished (Weinmann, 2013), and the suitability of a feature may strongly depend on the particular application.

In the following, we address feature extraction in more detail and describe methods for automatically detecting and describing features in either images (Section 3.1) or 3D point cloud data (Section 3.2). As modern devices offer the opportunity to acquire 2D and 3D image sequences with high frame rates, the extension to the spatio-temporal domain represents a further topic of interest (Section 3.3). Finally, summarising the main ideas, concluding remarks are provided (Section 3.4).

3.1 Feature Extraction from 2D Imagery

In general, images provide visual information represented on a regular 2D grid. Consequently, the aim of image processing techniques is to process and analyse images in order to describe and understand their content. For this purpose, a large variety of visual features have been introduced, and many investigations have been carried out that clearly demonstrate that such features significantly alleviate common tasks in photogrammetry, remote sensing, computer vision and robotics. In principle, a *visual feature* represents a piece of information that represents either a local or global property of an image as a whole or an object within the image. A recent survey on visual features (Weinmann, 2013) focuses on the categorisation of different types of visual features (Figure 3.1), the link of all these feature types to a common source in visual perception, and the extension from single images to image sequences.

In the following, the main ideas of the different feature types are briefly considered, and a selection of state-of-the-art approaches for feature extraction is provided. A complete survey is beyond the scope of this chapter, so only the most widely used approaches with a significant impact on recent research are discussed.

3.1.1 Pixel Attributes

The most obvious type of visual feature is based on dividing an image into the smallest possible elements. Since the introduction of digital cameras, digital images can be considered as discrete 2D projections of intensity measurements, which are arranged on a regular image raster. Each point on this regular image raster is referred to as a picture element (pixel). Depending on the camera type, the captured intensity information can be represented as grey-valued images where each pixel is assigned an intensity value, or as colour images where each

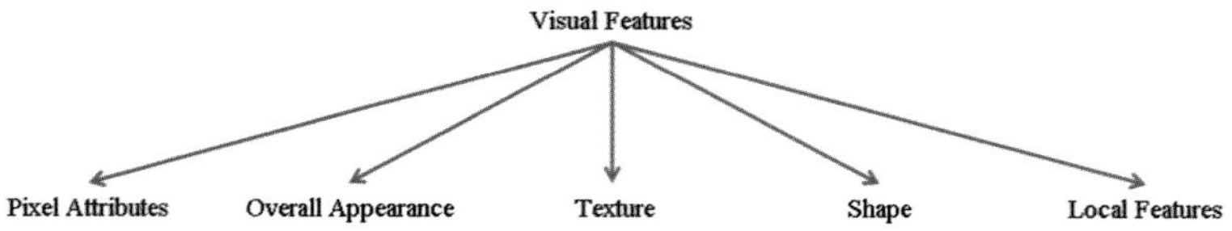

Figure 3.1 Different types of visual features contained in images.

pixel is assigned three values corresponding to the particular intensity in the red, green and blue channels. Such intensity information represents a perceptual attribute that can help to detect and distinguish objects in an image or image regions with different semantic content. Consequently, *intensity* represents a visual feature.

In recent decades, the availability of various other sensor types that capture information with a matrix of sensing elements has grown rapidly. Thus, a large variety of information, even beyond the visible spectrum (e.g. thermal infrared information or information from numerous spectral bands), can be transferred to 2D image representations. This is exploited by thermal cameras and satellite systems, for instance. In particular, satellite systems used in remote sensing for high-resolution satellite Earth observation exploit multi-spectral sensors to capture information in different spectral bands (Weidner, 2005), for example four spectral bands for IKONOS and QuickBird (blue, green, red and near infrared) or eight spectral bands for Landsat-7 (covering the visible, near infrared, short-wave infrared and thermal infrared spectra). The data can be used to estimate land cover/use and for vegetation monitoring. Additionally, hyperspectral sensors with a significantly higher number of channels with narrow bandwidth are now available, for example with 36 spectral bands for MODIS or 244 spectral bands for the planned EnMAP satellites. Such sensors providing high spectral resolution are often used to estimate the geochemical and biochemical composition of the atmosphere (Richards, 2005).

For other types of information that may be visualised as images (e.g. information in the form of range images or depth images), intensity information may not only arise from measured spectral information, but also from a representation of data associated to each pixel and thus possibly even abstract information. For this representation, an appropriate mapping function has to be defined or derived. A selection of images visualising different types of information are provided in Figure 3.2.

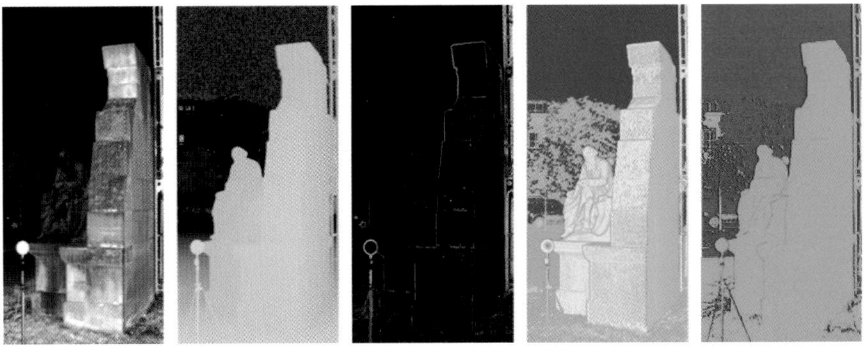

Figure 3.2 Images with different types of information: intensity image, range image, edge image, image encoding the local variance of range values, and binary image based on thresholding the local variance of range values.

3.1.2 Overall Appearance

Other techniques for extracting visual features focus on describing the global properties of an image and thus derive a holistic representation. Basic features are taken from the mean value or standard deviation of intensity values, and also from the respective skewness, kurtosis or entropy. To describe further properties of an image, the distribution of intensity values may be exploited. This distribution can be described via histograms, which represent the occurrence of single intensity values and discard their spatial arrangement. A consideration of the spatial arrangement of intensity values may involve moment theory (Flusser et al., 2009; Hu, 1962; Kotoulas and Andreadis, 2005). By exploiting moments it is possible to derive moment invariants, such as Zernike moment invariants (Teague, 1980; Teh and Chin, 1988) or orthogonal Fourier–Mellin moments (Sheng and Shen, 1994), which are invariant to certain mathematical transformations of an image (e.g. scaling or rotation). Those moments with a lower order capture the low-frequency information of an image (i.e. the coarse structure of the image function), whereas moments of higher order capture high-frequency information (i.e. image details). Thus, an image can be represented as a superposition of such moments. Alternatively, approaches based on the discrete Fourier transform (DFT) (Nixon and Aguado, 2008) or the discrete cosine transform (DCT) (Ahmed et al., 1974) also preserve the spatial arrangement of intensity values.

3.1.3 Texture

Not only intensity values, but also their spatial arrangement, may help to separate local image regions corresponding to different objects or different semantic content. This spatial arrangement may contain image regions of more or less homogeneity, which is often indicated by spatially repetitive (micro)structures across variations of scale and orientation. Consequently, the level of homogeneity that is commonly referred to as *texture* may be considered a visual feature. In general, texture can be defined as a function of the spatial variation of intensity values within an image (Chen et al., 1998). This definition encapsulates contextual information involving intensity values within a local neighbourhood, their spatial arrangement and different scales of resolution. In principle, approaches for describing texture can be categorised into three different groups (Haralick, 1979; Van Gool et al., 1985; Zhang and Tan, 2002): (1) statistical approaches, which describe texture via the statistical distribution of intensity values, (2) structural approaches, which describe texture as composition of texture primitives or textural elements (texels) that occur repeatedly with respect to certain placement rules, and (3) model-based approaches, which exploit probability models or a set of basis functions to describe textural image properties.

Prominent approaches exploiting a set of basis functions are based on the use of wavelet functions, such as Haar-like features (Papageorgiou et al., 1998; Viola and Jones, 2001), which approximate Haar wavelets with adjacent rectangular

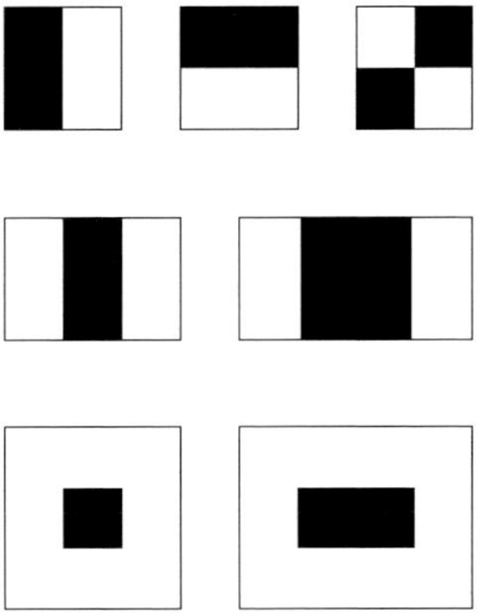

Figure 3.3 The structure of basic filter masks used for calculating Haar-like features. Two different weights are used to weight the contribution of pixels within the black and white regions. Modifications of these filter masks can easily be defined by rotation, scaling or inversion.

regions (Figure 3.3). Within each region, the respective intensity values are summed. Finally, depending on the approximated Haar wavelet, a weighted sum of the resulting values for these regions is calculated, where the weights can be positive or negative. The rectangular approximation allows the use of integral images and therefore Haar-like features can be calculated very efficiently. Other wavelet functions that are commonly used for deriving texture features are Gabor wavelets (Turner, 1986). These spatial filters are represented by sinusoidal plane waves within elliptical 2D Gaussian envelopes and, typically, a bank of Gabor filters of different scale and orientation is applied to analyse texture.

3.1.4 Shape

Early approaches to feature extraction revealed that information is not uniformly distributed over the whole image (Gibson, 1950; Attneave, 1954). Instead, information is concentrated along contours and particularly at points on a contour where its direction changes (i.e. corners). Such a contour typically arises from an abrupt change in intensity or texture that is likely to correspond to different materials, varying surface orientation or discontinuities with respect to illumination, distance or reflectance behaviour.

In general, approaches for describing 2D shapes can be categorised into either contour-based approaches or region-based approaches (Zhang and Lu, 2004). Contour-based approaches either exploit the global properties of a contour (e.g. perimeter, compactness, eccentricity, shape signature, Fourier descriptors or wavelet descriptors) or structural properties (e.g. chain codes, polygonal approximations, splines or boundary moments). Instead of only focusing on contours, region-based approaches also account for the region within a closed contour. Consequently, they may either address global properties (e.g. size, orientation, area, eccentricity or moments) or structural properties (e.g. convex hull, media axis or core).

In order to detect contours within an image, standard approaches are based on the convolution of the image with a filter mask. Suitable filter masks have been proposed with the Roberts operator (Roberts, 1965), the Prewitt operator (Prewitt and Mendelsohn, 1966) and the Sobel operator (Sobel, 1970). The Sobel operator in particular has become popular, as it combines both smoothing and gradient estimation. A more sophisticated approach to contour detection has been proposed with the Canny operator (Canny, 1981). The Canny operator is based on convolving the image with the Sobel operators in the x- and y-directions. This reveals the partial derivatives and thus the gradient of the image function at each pixel in the horizontal and vertical directions in the form of gradient images. From these gradient images, the corresponding gradient magnitude and gradient direction of each pixel are derived. To locate edges and thus contours, a non-maximum suppression is carried out to obtain reliable contour information. Based on these contours, the respective edges are traced through the image by hysteresis thresholding. As a result, reliable and often closed contours are available.

Furthermore, it may be taken into account that specific contours can be assigned a contour parameterisation by exploiting a mathematical model. Accordingly, it is possible to locate shapes in an image by involving *a priori* knowledge on the expected contours. To detect straight lines, the Hough transform (Hough, 1962) has been presented, where each point on a contour votes for bins along a curve in a discrete parameter space also referred to as accumulator or Hough space. The dimension of the accumulator is represented by the number of unknown parameters, and the respective curves may intersect. Maxima in the accumulator indicate sets of parameters that are supported by several points on the contour and thus suited to approximate the respective contour. By adapting the model, the Hough transform has been transferred to circle and ellipse detection. The extension to arbitrary shapes has been proposed with the generalized Hough transform (GHT) (Ballard, 1981).

3.1.5 Local Features

Early investigations on feature extraction revealed that highly informative image points are located at corners or blobs in an image (Attneave, 1954). Such features represent local image characteristics and are therefore referred to as *local features*. In general, a local feature represents a pattern varying from its spatial neighbourhood (Tuytelaars and Mikolajczyk, 2008); the variation may be due to a change in one or more image properties (e.g. colour, intensity or texture). Accordingly, it is possible to define corners, blobs, specific edge elements (edgels) or small distinctive image regions as local features. Such local features share very interesting properties. They can be localised accurately in the image and remain stable over time. Taking into consideration their local image neighbourhood, even an individual identification of such features is possible. Furthermore, local features can be extracted very efficiently. Due to these properties, local features have become very popular in recent years, and they have been introduced for a wide range of applications such as object recognition, autonomous navigation and exploration, image and video retrieval, image registration, 3D reconstruction or scene interpretation. Recent surveys show in great detail how such features can be detected, described and finally recognised in a similar image again (Tuytelaars and Mikolajczyk, 2008; Li and Allinson, 2008; Weinmann, 2013).

Typically, approaches for extracting local features can be split into *feature detection* to adequately extract the respective image locations and *feature description* to appropriately represent the local image characteristics. A feature detector thus only yields distinctive image locations (also referred to as *interest points* or *keypoints*) without a feature representation, and the desired representation is subsequently derived with a feature descriptor. Consequently, the tasks of feature detection and feature description can be treated separately, and a large variety of approaches have been presented in the literature for both tasks. In the following, a small selection of the most commonly used approaches are presented. Further alternatives and more details can be found in recent survey papers (Tuytelaars and Mikolajczyk, 2008; Li and Allinson, 2008; Weinmann, 2013).

3.1.6 Local Feature Detection

Considering local features in the form of corners, blob-like structures or distinctive image regions that often represent characteristic image locations, the detection of local features in an image may be based on different principles. We hence consider particular strategies for detecting respective types of local features separately.

3.1.6.1 Corner Detectors

Methods for detecting corners in an image were introduced with the Moravec corner detector (Moravec, 1977). An improvement to this approach resulted in the Harris corner detector (Harris and Stephens, 1988) which is based on the *structure tensor* **S**, which encapsulates the local statistics of the image function. Formally, the structure tensor can be described as second-moment matrix with

$$\mathbf{S} = \sum_{(x,y)\in\mathcal{N}} w(x,y) \begin{bmatrix} \mathbf{I}_x^2 & \mathbf{I}_x\,\mathbf{I}_y \\ \mathbf{I}_x\,\mathbf{I}_y & \mathbf{I}_y^2 \end{bmatrix} \tag{3.1}$$

where only pixels (x, y) within a local neighbourhood $\mathcal{N}$ are considered and $w(x, y)$ represents a window function. As window function, either a uniform weighting function or an isotropic Gaussian weighting function is typically chosen. Expressions $\mathbf{I}_x$ and $\mathbf{I}_y$ represent the partial derivatives of the image function

$\mathbf{I}(x, y)$ in the x- and y-directions, respectively. More specifically, the consideration of the eigenvalues λ_1 and λ_2 of the structure tensor with $\lambda_1 \geq \lambda_2 \geq 0$ allows us to categorise pixels into (1) pixels within homogeneous image regions, (2) pixels on edges and (3) pixels representing corners:

- Two small eigenvalues λ_1 and λ_2 indicate a local image neighbourhood with approximately constant intensity.
- The combination of a high eigenvalue λ_1 and a low eigenvalue λ_2 indicates an edge.
- Two high eigenvalues λ_1 and λ_2 indicate a corner.

A particular categorisation based on a single value has been proposed with a measure commonly referred to as cornerness c with

$$c = \det(\mathbf{S}) - \kappa \operatorname{tr}^2(\mathbf{S}) = \lambda_1 \lambda_2 - \kappa (\lambda_1 + \lambda_2)^2 \tag{3.2}$$

where κ represents a tunable sensitivity parameter, typically chosen between 0.04 and 0.06 (Harris and Stephens, 1988; Kerr et al., 2008). Small eigenvalues λ_1 and λ_2 cause relatively small values for the first and second terms. Thus, the cornerness reaches a small absolute value that is characteristic of a local image neighbourhood with approximately constant intensity. If λ_1 is high and λ_2 is low, the second term becomes dominant and the cornerness is negative. This reveals an edge. For two high eigenvalues λ_1 and λ_2, the first term becomes dominant in comparison to the second term. Thus, the cornerness is positive and reveals a corner. By additionally accounting for changes in image scale, a multi-scale Harris detector has been proposed (Dufournaud et al., 2000; Mikolajczyk and Schmid, 2001). Further extensions have been presented with the Harris-Laplace detector (Mikolajczyk and Schmid, 2001) and the Harris-affine detector (Mikolajczyk and Schmid, 2004). Based on the structure tensor, its eigenvalues and a measure describing the energy of the image function at the particular image location, the Förstner detector (Förstner and Gülch, 1987) has been proposed. This detector provides a higher localisation accuracy for interest points, but it also requires a higher computational effort.

A different strategy for detecting corners has been proposed with a features from accelerated segment test (FAST) detector (Rosten and Drummond, 2005). The FAST detector relies only on a small part of the local image neighbourhood to recover corners. Specifically, for each pixel (x, y), the FAST detector accounts only for those 16 pixels on a surrounding discretised circle, a Bresenham circle of radius $r = 3$ (Figure 3.4) and performs a test by comparing the intensity values of pixels on the Bresenham circle to the intensity value $\mathbf{I}(x, y)$. For efficiency, the comparison is performed with an accelerated segment test (AST) according to which the pixel (x, y) represents a potential corner if at least $N = 12$ contiguous pixels on the Bresenham circle are brighter than $\mathbf{I}(x, y) + t$ or darker than $\mathbf{I}(x, y) - t$, where t represents a threshold. Accordingly, a larger

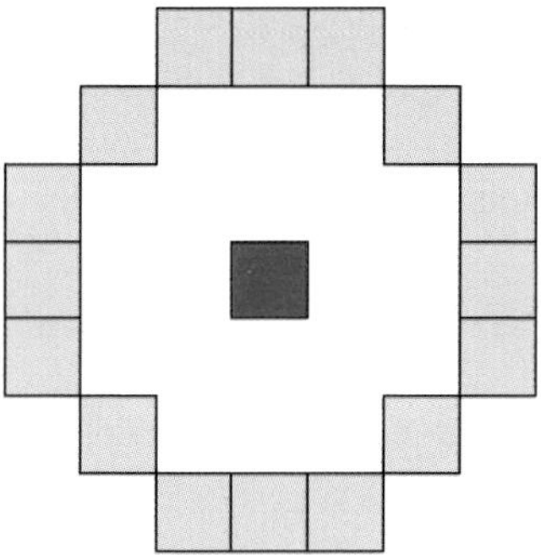

Figure 3.4 The structure of a Bresenham circle of radius $r = 3$. The centre pixel is highlighted in red.

threshold t yields few but stronger interest point candidates, whereas a smaller threshold t also results in interest point candidates with smoother gradients. A subsequent non-maximum suppression finally yields the particular interest points. Due to the efficient methodology, the FAST detector remains one of the fastest approaches for detecting local features. As the FAST detector is not scale-invariant, a multi-scale FAST detector has been proposed that involves a scale selection based on the Laplacian function (Lepetit and Fua, 2006). Further extensions accounting for scale- and rotation-invariant feature detection have also been proposed (Rublee et al., 2011; Leutenegger et al., 2011).

3.1.6.2 Blob Detectors

Methods for detecting blob-like image structures can be traced back to the Hessian detector (Beaudet, 1978) exploiting the Hessian matrix **H** with

$$\mathbf{H} = \begin{bmatrix} \mathbf{I}_{xx} & \mathbf{I}_{xy} \\ \mathbf{I}_{xy} & \mathbf{I}_{yy} \end{bmatrix} \tag{3.3}$$

which consists of the second-order image derivatives of the image function $\mathbf{I}(x, y)$. From the Hessian matrix, two measures are derived in order to detect blob-like image structures: its determinant, also referred to as the *discriminant*, and its trace, also referred to as the *Laplacian*. Respective interest points can be detected at locations where these measures reach a local maximum above a certain predefined threshold. Detectors based on the determinant of the Hessian matrix are often referred to as determinant-of-Hessian (DoH) detectors. As the Hessian detector is only invariant to image rotation, an extension to the multi-scale Hessian detector (Deriche and Giraudon, 1993) has been proposed that also accounts for scale-invariant feature detection. In analogy to approaches for corner detection, further modifications have been presented with the Hessian-Laplace detector (Mikolajczyk and Schmid, 2001) and the Hessian-affine detector (Mikolajczyk et al., 2005).

Focusing on the Laplacian function as a measure to detect blob-like image structures, an efficient approximation can be realised by applying a difference-of-Gaussian (DoG) function. Consequently, the particular interest point detector is commonly referred to as a DoG detector. To additionally account for multi-scale representation, a multi-scale DoG detector has been proposed that also represents the first part of the scale invariant feature transform (SIFT) and is hence also referred to as a SIFT detector (Lowe, 1999, 2004). The SIFT detector consists of two successive steps:

- *Scale-space extrema detection.* To provide scale-invariant features, a scale-space representation has to be derived. For this purpose, the image function $\mathbf{I}$ is first convolved with Gaussian kernels $G(x, y, \sigma_i)$ of variable scale σ_i:
$$L(x, y, \sigma_i) = G(x, y, \sigma_i) * \mathbf{I}(x, y) \tag{3.4}$$
Subsequently, one of the resulting Gaussian-smoothed images is reduced in size by subsampling, and again convolved with Gaussian kernels of variable scale. By further repetitions of this procedure, a grouping of the Gaussian-smoothed images according to their size, and a sorting with respect to their scale, the Gaussian scale-space is constructed, which is also referred to as the Gaussian pyramid. A subsequent subtraction of neighbouring Gaussian-smoothed images of the same size yields the scale-space of Gaussian differences, which is also known as a DoG pyramid:
$$S(x, y, \sigma_i) = \left[G(x, y, \sigma_i) - G(x, y, \sigma_{i-1})\right] * \mathbf{I}(x, y) \tag{3.5}$$
$$= L(x, y, \sigma_i) - L(x, y, \sigma_{i-1}) \tag{3.6}$$
By considering all DoG images of a specific size and comparing each sample point to other points within a local $(3 \times 3 \times 3)$-neighbourhood (i.e. a comparison to its eight neighbours at the same scale as well as its nine neighbours at both neighbouring scales), interest point candidates can be determined as local extrema.
- *Localisation of interest points.* Once interest point candidates have been detected, the next step consists of improving their location to subpixel accuracy and removing candidates with low stability. To estimate the subpixel-accurate image location of each interest point candidate, a 3D quadratic function is fitted to neighbouring points with respect to image location and scale in order to determine the interpolated location of the extremum (Brown and Lowe, 2002). This is done by approximating the scale-space function by a Taylor expansion up to quadratic terms. To remove unstable points, two criteria are addressed: (1) points with low contrast are sensitive to noise and can easily be discarded by thresholding; (2) points located along edges can hardly be distinguished and are discarded by considering the Hessian matrix computed at the location

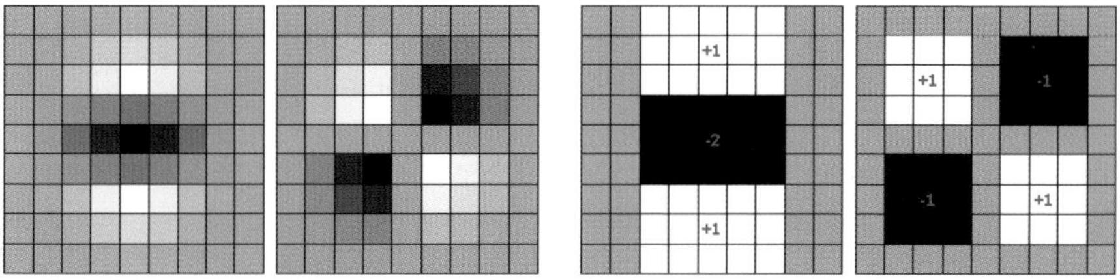

Figure 3.5 Gaussian second-order partial derivatives and their box filter approximations.

and scale of the particular interest point candidate, where the respective second-order derivatives are derived from the neighbouring points. From the determinant and the trace of the Hessian matrix, a measure is derived that reveals if the ratio of principal curvatures is above a certain threshold and the respective interest point candidate has to be removed.

Furthermore, it has been proposed to approximate the scale-space representation of the Hessian matrix by applying box filters. This results in a speeded up robust feature (SURF) detector (Bay et al., 2006). In general, a box filter subdivides a window into rectangular areas and provides an equal weight for the contribution of all pixels within these areas (Figure 3.5). Accordingly, integral images can be used to efficiently calculate the sum of all intensity values within any upright rectangular area, independent of its size. In analogy to the multi-scale Hessian detector, interest point candidates are detected at locations in the scale-space where the determinant of the approximated Hessian matrix reaches a maximum within a local $(3 \times 3 \times 3)$-neighbourhood. Furthermore, the location of resulting interest points is improved to subpixel accuracy in a way similar to the SIFT detector.

3.1.6.3 Region Detectors

To detect distinctive image regions, a method has been presented that focuses on the extraction of maximally stable extremal regions (MSERs) and is therefore referred to as the MSER detector (Matas et al., 2002). The MSER detector is based on (1) sorting all pixels by intensity, (2) placing them in the image in either increasing or decreasing order, and (3) recovering the connected components with the respective areas by applying the union-find algorithm (Sedgewick, 1988). Resulting from this, a data structure is available where the area of each connected component is stored as a function of intensity, and MSERs correspond to thresholds for which the relative area change is extremal. Accordingly, MSERs can be considered as connected components of an appropriately thresholded image (Tuytelaars and Mikolajczyk, 2008), where each MSER is either brighter or darker than its surrounding area. As the feature matching process should later result in point correspondences, each MSER is assigned its centroid and, optionally, also the respective threshold value. A further optimisation of the MSER detector has been proposed by replacing the union-find algorithm with a more efficient strategy and thus reducing computational effort as well as memory consumption (Nister and Stewenius, 2008). Furthermore, an extension involving a scale-space representation has been presented with the multi-resolution MSER detector (Forssen and Lowe, 2007).

3.1.7 Local Feature Description

To find a similar image location in a different image again, early approaches were typically based on correlating small image patches of the local neighbourhoods around the detected distinctive image locations. However, this is very time-consuming when comparing a large number of features, particularly when also considering different scales and rotations between the respective image patches. Consequently, a compact representation is desirable that allows a more efficient comparison. For this reason, most of the commonly used approaches exploit descriptors representing local characteristics in the form of vectors and thus allowing a comparison via standard distance metrics such as the Euclidean distance or a general Minkowski metric in order to efficiently derive feature correspondences between different images.

A powerful vector representation was introduced with the SIFT descriptor (Lowe, 1999, 2004), which was inspired from investigations on biological vision, and which represents the description part of the SIFT. The SIFT descriptor takes into account that similar features may provide the relevant information around the respective interest points at different scales and different orientations. As a consequence, the resulting SIFT descriptors are invariant to changes in image scale and image rotation. Furthermore, the SIFT descriptor is designed to be robust to image noise, changes in illumination and small changes in viewpoint. This is achieved by exploiting orientation histograms and, according to (Lowe, 1999, 2004), motivated by investigations revealing that complex neurons in the primary visual cortex respond to a gradient at a particular orientation and spatial frequency, but the location of the gradient on the retina is allowed to shift over a small receptive field rather than being precisely localised (Edelman et al., 1997). These seminal investigations have already demonstrated that a feature matching thus results in better persistence of feature correspondences under 3D rotation of objects, which improves object recognition from different viewpoints. In order to achieve these beneficial properties, the construction of a SIFT descriptor consists of two steps:

- *Orientation assignment.* In order to later provide a descriptor with rotational invariance, the dominant orientation of an interest point is derived, according to which the descriptor calculation is aligned. For this purpose, the closest Gaussian-smoothed image is considered, and an orientation histogram with 36 angular bins representing sections of $10°$ is sampled

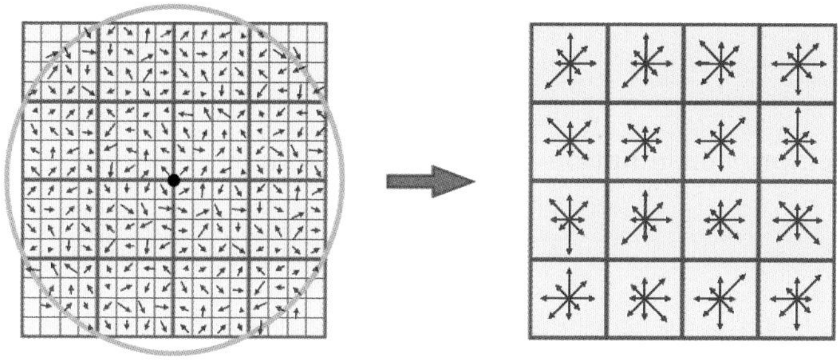

Figure 3.6 Structure of the SIFT descriptor: the local neighbourhood of a keypoint is split into 4 × 4 subregions, and the gradient information within each subregion is used to derive an orientation histogram.

from the orientations of the local image gradients. The contribution of each gradient orientation is weighted by the respective gradient magnitude and a Gaussian window function centred on the interest point. The Gaussian window function is used for weighting the contribution of local gradients around an interest point with respect to their distance. A larger distance thus corresponds to a lower weight. In the orientation histogram, more dominant directions of the local gradients result in peaks. Consequently, the dominant orientation assigned to the respective interest point is represented by the highest peak. For each other peak above 80% of this peak, an additional interest point with the same image location and the same scale is generated and assigned the respective orientation. A final interpolation of the considered peaks in the orientation histogram yields a refined and more accurate orientation assignment.

- *Generation of keypoint descriptors.* Once an interest point has been assigned its dominant orientation, image gradients within the local neighbourhood in the closest Gaussian-smoothed image are rotated with respect to this dominant orientation. Subsequently, the aligned neighbourhood is divided into 4 × 4 subregions (Figure 3.6). For each subregion, the local gradient orientations are accumulated and an orientation histogram with eight angular bins representing sections of 45° is derived. At this stage, the contribution of each gradient orientation is again weighted by the respective gradient magnitude and a Gaussian window function centred on the interest point. The concatenation of the histogram bins across all 16 subregions yields a descriptor in the form of a vector with 128 elements. Finally, a normalisation is carried out in order to increase the robustness to changes in illumination, and this finally yields the respective SIFT descriptor.

Considering the fact that some SIFT features extracted from an image will possibly not have any correct match in a different image, some of the detected

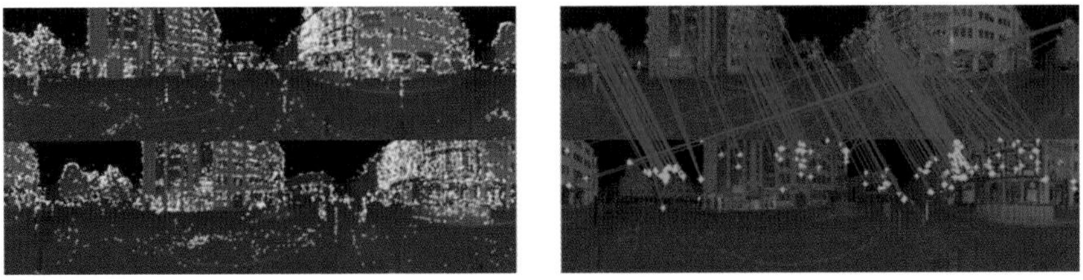

Figure 3.7 Comparison of two images. The SIFT detector is used to find distinctive image locations (yellow), and the subsequent matching based on the respective SIFT descriptors results in reliable 2D/2D correspondences between locations in the first image (red) and locations in the second image (green). Note that the correspondences not following the main trend are not erroneous, because the images cover 360° in the horizontal direction.

feature correspondences will have to be discarded in the matching process. However, applying a global threshold on the Euclidean distance to the most similar descriptor is not suitable, as some descriptors might be much more discriminative than others. Hence, a ratio test has been proposed (Lowe, 2004) that compares the Euclidean distances of a keypoint descriptor in the first image to the nearest neighbour and to the second-nearest neighbour in the second image. This ratio $r \in [0, 1]$ of these distances has to be below a certain threshold t in order to obtain distinctive features and thus correspondences with a higher reliability. A threshold of $t = 0.8$, for instance, indicates that the distance to the nearest neighbour has to be smaller or equal to 80% of the distance to the second-nearest neighbor. In general, a low value of r indicates a high similarity to only one of the derived descriptors belonging to the other image. Thus, the ratio r describes the distinctiveness of corresponding features, and meaningful feature correspondences arise from a greater difference of the considered distances. This matching procedure significantly reduces the occurrence of ambiguities, and it supports the idea that more distinctive features are more likely to be correctly matched than less distinctive features. Figure 3.7 provides an example for detecting distinctive image locations via the SIFT detector and deriving feature correspondences via matching based on the respective SIFT descriptors.

A feature descriptor that is closely related to the SIFT descriptor has been presented with the histogram of oriented gradient (HOG) descriptor (Dalal and Triggs, 2005). The HOG descriptor also relies on orientation histograms constructed from the gradient orientations within the local neighbourhood of an interest point. Specifically, the local neighbourhood is divided into subregions. For all pixels within a subregion, the gradient orientations are derived and quantised into an orientation histogram with a fixed number of angular bins. The concatenation of the histogram entries for all subregions and a subsequent normalisation finally yield the respective HOG descriptor assigned to the interest point.

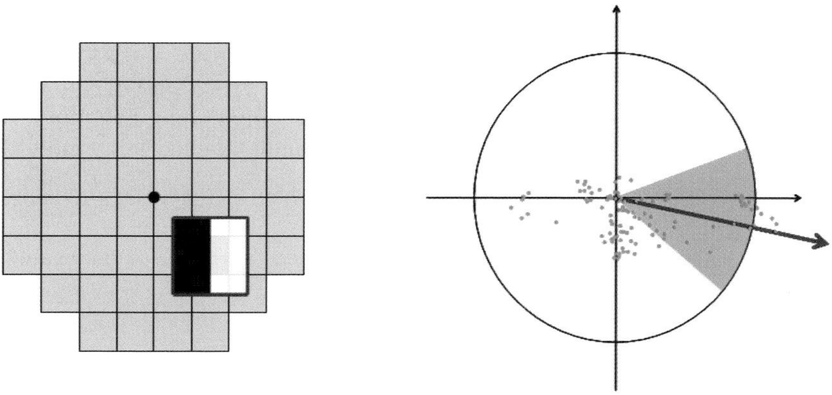

Figure 3.8 Deriving the dominant orientation for the SURF descriptor. For each pixel in the local neighbourhood, the Haar wavelet responses in the x- and y-directions are represented in a coordinate frame. A sliding window technique is then exploited to find the longest vector resulting from the respective Haar wavelet responses.

A further descriptor related to the SIFT descriptor has been presented with the speeded up robust feature (SURF) descriptor (Bay et al., 2006). A SURF descriptor also focuses on describing the local neighbourhood of an interest point relative to its dominant orientation. Hence, to derive a SURF descriptor, the following two steps are performed:

- *Orientation assignment.* To derive the dominant orientation of an interest point, the Haar wavelet responses in the x- and y-directions within a circular neighbourhood around the interest point are exploited. These wavelet responses are weighted by a Gaussian window function centred on the interest point in order to account for the relative distance to the interest point. Subsequently, the weighted wavelet responses are represented as points in a 2D coordinate frame spanned by the response strengths in the x- and y-directions (Figure 3.8). Accumulating all points within a sliding orientation window covering $60°$ yields a local orientation vector. The orientation of the sliding window for which the local orientation vector reaches a maximum length is defined as the dominant orientation.
- *Generation of keypoint descriptors.* After aligning the local neighbourhood with respect to the dominant orientation, the local neighbourhood is divided into 4×4 subregions. For each subregion, the Haar wavelet responses are derived with respect to the local x- and y-directions and weighted by a Gaussian window function centred on the interest point. Subsequently, four characteristic measures are derived within each subregion: (1) the sum over the weighted wavelet responses in the local x-direction, (2) the sum over the weighted wavelet responses in the local y-direction, (3) the sum over the weighted absolute wavelet responses in the x-direction, and (4) the sum over the weighted absolute wavelet

responses in the y-direction. The concatenation of these four values for all 16 subregions results in a 64-dimensional vector representation. A subsequent normalisation increases the robustness to changes in illumination and finally yields the respective SURF descriptor.

A matching process relying on SURF descriptors can be carried out in analogy to the matching of SIFT descriptors.

Applying a different strategy to also exploit orientation histograms based on gradient information within the local neighbourhood, a further alternative for describing interest points has been proposed with the DAISY descriptor (Tola et al., 2008, 2010). Hence, when compared to the SIFT-like descriptors, the DAISY descriptor exhibits similar properties with respect to changes in image scale, image rotation, image noise, illumination and viewpoint. In contrast to the SIFT descriptor, which has been introduced for sparse wide-baseline matching, the DAISY descriptor can be computed in a more efficient way and thus even allows a dense depth map estimation from wide-baseline image pairs. In contrast to the SURF descriptor, which can be derived more efficiently than the SIFT descriptor, the SIFT and DAISY descriptors show a better matching performance, because the SURF descriptor only relies on equally contributing gradient orientations because of a missing spatial weighting. Consequently, the DAISY descriptor provides both computational efficiency and reliable feature correspondences. Similar to the SIFT descriptor, the standard DAISY descriptor is based on convolving the image function $\mathbf{I}(x, y)$ with Gaussian kernels. This convolution introduces a smoothing and thus reduces the influence of noise on the descriptor. Specifically, an operator $(\,\cdot\,)^+$ with $(\xi)^+ = \max\{\xi, 0\}$ is defined, and orientation maps G_o with

$$G_o = \left(\frac{\partial \mathbf{I}(x, y)}{\partial o}\right)^+ \tag{3.7}$$

are derived for eight quantised directions differing by 45°, where o is the orientation of the derivative and $G_o(u, v)$ represents the image gradient norm at location (u, v). The subsequent convolution of each orientation map with several Gaussian kernels G_{σ_i} of different scale σ_i results in convolved orientation maps $G_o^{\sigma_i}$ with

$$G_o^{\sigma_i} = G_{\sigma_i} * \left(\frac{\partial \mathbf{I}(x, y)}{\partial o}\right)^+ \tag{3.8}$$

where the applied Gaussian filters are separable. Therefore, convolutions with a Gaussian kernel of large scale may be derived from successive convolutions with smaller Gaussian kernels. For each of the 25 positions (u, v) on concentric circles around the interest point at (u_0, v_0) (Figure 3.9), an orientation histogram $\mathbf{h}_{\sigma_i}(u, v)$ with eight bins and

$$\mathbf{h}_{\sigma_i}(u, v) = \left[G_1^{\sigma_i}(u, v), \ldots, G_8^{\sigma_i}(u, v)\right]^T \tag{3.9}$$

The adaptation of feature descriptors to the spatio-temporal domain has also been addressed in recent years. A popular descriptor has been presented with the spatio-temporal version of the SIFT descriptor (Cheung and Hamarneh, 2007; Scovanner et al., 2007) based on orientation histograms. Similarly, histograms of gradient orientations are also exploited by the HOG3D descriptor (Kläser et al., 2008), which represents a spatio-temporal descriptor extending the standard HOG descriptor. Alternatively, an extension of the DAISY descriptor to the spatio-temporal domain by means of optical flow priors has been presented (Trulls et al., 2012), which is based on capturing the changes around an interest point in time instead of trying to describe the respective spatio-temporal volume.

In analogy to local 2D, 2.5D and 3D features, comparisons of numerous approaches for extracting spatio-temporal features have been presented. As a result, it can be stated that the performance of the different approaches strongly depends on the particular application, and there is not yet a clear trend towards a *best* feature detector or a *best* feature descriptor. For example, recent evaluations of different approaches for detecting and describing spatio-temporal features address action recognition (Wang et al., 2009) as well as video retrieval and categorisation (Stöttinger et al., 2011).

3.4 Conclusions

In this chapter, we have addressed the issue of feature extraction from 2D imagery and 3D point clouds. This issue has become the basis and core of a plethora of applications in photogrammetry, remote sensing, computer vision and robotics. The intention of this chapter is to provide a rough overview of 2D and 3D feature extraction. Consequently, we have reflected the main ideas of different types of visual features as well as different types of point cloud features, and have discussed recent extensions to the spatio-temporal domain. The focus has thus been set on a small selection of the most widely used approaches with significant impact on recent research. In general, the choice of adequate features still strongly depends on the particular application, but powerful methods are available among all feature types. Because – for both 2D imagery and 3D point cloud data – local features provide interesting characteristics with respect to efficiency, accurate localisation, stability over time and individual identification, these features are nowadays used for a variety of applications from extracting, describing and comparing objects to tasks focusing on image or point cloud registration, data retrieval, scene analysis and the autonomous navigation of vehicles.

References

Ahmed, N., Natarajan, T., and Rao, K. (1974). Discrete cosine transform. *IEEE Transactions on Computers*, C-23(1):90–93.

Al-Manasir, K. and Fraser, C. (2006). Registration of terrestrial laser scanner data using imagery. *The Photogrammetric Record*, 21(115):255–268.

Alexandre, L. (2012). 3D descriptors for object and category recognition: a comparative evaluation. In *Proceedings of the IROS Workshop on Color-Depth Camera Fusion in Robotics*.

Allaire, S., Kim, J., Breen, S., Jaffray, D., and Pekar, V. (2008). Full orientation invariance and improved feature selectivity of 3D SIFT with application to medical image analysis. In *Proceedings of the IEEE Computer Society Conference on Computer Vision and Pattern Recognition Workshops*, pages 1–8.

Attneave, F. (1954). Some informational aspects of visual perception. *Psychological Review*, 61(3):183–193.

Ballard, D. (1981). Generalizing the Hough transform to detect arbitrary shapes. *Pattern Recognition*, 13(2):111–122.

Bay, H., Tuytelaars, T., and Van Gool, L. (2006). SURF: speeded up robust features. In Leonardis, A., Bischof, H., and Pinz, A., editors, *ECCV 2006, Part I*, pages 404–417. LNCS, Vol. 3951, Springer, Heidelberg, Germany.

Beaudet, P. (1978). Rotationally invariant image operators. In *Proceedings of the International Joint Conference on Pattern Recognition*, pages 579–583.

Brown, M. and Lowe, D. (2002). Invariant features from interest point groups. In *Proceedings of the British Machine Vision Conference*, pages 656–665.

Canny, J. (1981). A computational approach to edge detection. *IEEE Transactions on Pattern Analysis and Machine Intelligence*, 8(6):679–698.

Chen, C., Pau, L., and Wang, P. (1998). *Handbook of pattern recognition and computer vision* (2nd edition). World Scientific Publishing, Singapore.

Cheung, W. and Hamarneh, G. (2007). n-SIFT: n-dimensional scale invariant feature transform for matching medical images. In *Proceedings of the IEEE International Symposium on Biomedical Imaging: From Nano to Macro*, pages 720–723.

Dahl, A., Aanæs, H., and Pedersen, K. (2011). Finding the best feature detector-descriptor combination. In *Proceedings of the Joint Conference on 3D Imaging, Modeling, Processing, Visualisation and Transmission*, pages 318–325.

Dalal, N. and Triggs, B. (2005). Histograms of oriented gradients for human detection. In *Proceedings of the IEEE Computer Society Conference on Computer Vision and Pattern Recognition*, Vol. 1, pages 886–893.

Demantké, J., Vallet, B., and Paparoditis, N. (2012). Streamed vertical rectangle detection in terrestrial laser scans for facade database production. *ISPRS Annals of the Photogrammetry, Remote Sensing and Spatial Information Sciences*, 1(3):99–104.

Deriche, R. and Giraudon, G. (1993). A computational approach for corner and vertex detection. *International Journal of Computer Vision*, 10(2):101–124.

Donoser, M. and Bischof, H. (2006). 3D segmentation by maximally stable volumes (MSVs). In *Proceedings of the International Conference on Pattern Recognition*, Vol. 1, pages 63–66.

Dufournaud, Y., Schmid, C., and Horaud, R. (2000). Matching images with different resolutions. In *Proceedings of the IEEE Computer Society Conference on Computer Vision and Pattern Recognition*, Vol. 1, pages 612–618.

Dutagaci, H., Cheung, C., and Godil, A. (2012). Evaluation of 3D interest point detection techniques via human-generated ground truth. *The Visual Computer*, 28(9):901–917.

Edelman, S., Intrator, N., and Poggio, T. (1997). Complex cells and object recognition. *Unpublished manuscript.*

Filin, S. and Pfeifer, N. (2005). Neighbourhood systems for airborne laser data. *Photogrammetric Engineering & Remote Sensing*, 71(6):743–755.

Flint, A., Dick, A., and Van Den Hengel, A. (2007). Thrift: local 3D structure recognition. In *Proceedings of the Biennial Conference of the Australian Pattern Recognition Society on Digital Image Computing Techniques and Applications*, pages 182–188.

Flitton, G., Breckon, T., and Megherbi, N. (2010). Object recognition using 3D SIFT in complex CT volumes. In *Proceedings of the British Machine Vision Conference*, pages 11.1–11.12.

Flitton, G., Breckon, T., and Megherbi, N. (2013). A comparison of 3D interest point descriptors with application to airport baggage object detection in complex CT imagery. *Pattern Recognition*, 46(9):2420–2436.

Flusser, J., Suk, T., and Zitova, B. (2009). *Moments and moment invariants in pattern recognition.* John Wiley & Sons, Chichester, UK.

Forssen, P.-E. and Lowe, D. (2007). Shape descriptors for maximally stable extremal regions. In *Proceedings of the IEEE International Conference on Computer Vision*, pages 1–8.

Förstner, W. and Gülch, E. (1987). A fast operator for detection and precise location of distinct points, corners and centres of circular features. In *Proceedings of the ISPRS Conference on Fast Processing of Photogrammetric Data*, pages 281–305.

Frome, A., Huber, D., Kolluri, R., Bülow, T., and Malik, J. (2004). Recognizing objects in range data using regional point descriptors. In Pajdla, T. and Matas, J., editors, *ECCV 2004, Part III*, pages 224–237. LNCS, Vol. 3023, Springer, Heidelberg, Germany.

Gibson, J. (1950). *The perception of the visual world.* Houghton Mifflin, Boston, USA.

Haralick, R. (1979). Statistical and structural approaches to texture. *Proceedings of the IEEE*, 67(5):786–804.

Harris, C. and Stephens, M. (1988). A combined corner and edge detector. In *Proceedings of The Alvey Vision Conference*, pages 147–151.

Hough, P. (1962). Method and means for recognizing complex patterns. *United States Patent 3069654.*

Hu, M.-K. (1962). Visual pattern recognition by moment invariants. *IRE Transactions on Information Theory*, 8(2):179–187.

Johnson, A. and Hebert, M. (1999). Using spin images for efficient object recognition in cluttered 3D scenes. *IEEE Transactions on Pattern Analysis and Machine Intelligence*, 21(5):433–449.

Jutzi, B. and Gross, H. (2009). Nearest neighbour classification on laser point clouds to gain object structures from buildings. *The International Archives of the Photogrammetry, Remote Sensing and Spatial Information Sciences*, 38(1-4-7/W5).

Kerr, D., Coleman, S., and Scotney, B. (2008). Comparing cornerness measures for interest point detection. In *Proceedings of the International Machine Vision and Image Processing Conference*, pages 105–110.

Kim, H. and Hilton, A. (2013). Evaluation of 3D feature descriptors for multi-modal data registration. In *Proceedings of the International Conference on 3D Vision*, pages 119–126.

Kläser, A., Marszalek, M., and Schmid, C. (2008). A spatio-temporal descriptor based on 3D-gradients. In *Proceedings of the British Machine Vision Conference*, pages 275:1–10.

Knopp, J., Prasad, M., Willems, G., Timofte, R., and Van Gool, L. (2010). Hough transform and 3D SURF for robust three dimensional classification. In Daniilidis, K., Maragos, P., and Paragios, N., editors, *ECCV 2010, Part VI*, pages 589–602. LNCS, Vol. 6316, Springer, Heidelberg, Germany.

Koelstra, S. and Patras, I. (2009). The FAST-3D spatio-temporal interest region detector. In *Proceedings of the Workshop on Image Analysis for Multimedia Interactive Services*, pages 242–245.

Kotoulas, L. and Andreadis, I. (2005). Image analysis using moments. In *Proceedings of the International Conference on Technology and Automation*, pages 360–364.

Laptev, I. (2005). On space-time interest points. *International Journal of Computer Vision*, 64(2-3):107–123.

Lari, Z. and Habib, A. (2012). Alternative methodologies for estimation of local point density index: moving towards adaptive lidar data processing. *The International Archives of the Photogrammetry, Remote Sensing and Spatial Information Sciences*, 39(B3):127–132.

Lee, I. and Schenk, T. (2002). Perceptual organization of 3D surface points. *The International Archives of the Photogrammetry, Remote Sensing and Spatial Information Sciences*, 34(3A):193–198.

Lepetit, V. and Fua, P. (2006). Keypoint recognition using randomized trees. *IEEE Transactions on Pattern Analysis and Machine Intelligence*, 28(9):1465–1479.

Leutenegger, S., Chli, M., and Siegwart, R. (2011). BRISK: binary robust invariant scalable keypoints. In *Proceedings of the IEEE International Conference on Computer Vision*, pages 2548–2555.

Li, J. and Allinson, N. (2008). A comprehensive review of current local features for computer vision. *Neurocomputing*, 71:1771–1787.

Linsen, L. and Prautzsch, H. (2001). Local versus global triangulations. In *Proceedings of Eurographics*, pages 257–263.

Lo, T. and Siebert, J. (2009). Local feature extraction and matching on range images: 2.5D SIFT. *Computer Vision and Image Understanding*, 113(12):1235–1250.

Lowe, D. (1999). Object recognition from local scale-invariant features. In *Proceedings of the IEEE International Conference on Computer Vision*, Vol. 2, pages 1150–1157.

Lowe, D. (2004). Distinctive image features from scale-invariant keypoints. *International Journal of Computer Vision*, 60(2):91–110.

Maas, H.-G. and Vosselman, G. (1999). Two algorithms for extracting building models from raw laser altimetry data. *ISPRS Journal of Photogrammetry and Remote Sensing*, 54(2–3):153–163.

Mallet, C., Bretar, F., Roux, M., Soergel, U., and Heipke, C. (2011). Relevance assessment of full-waveform lidar data for urban area classification. *ISPRS Journal of Photogrammetry and Remote Sensing*, 66(6):S71–S84.

Matas, J., Chum, O., Urban, M., and Pajdla, T. (2002). Robust wide baseline stereo from maximally stable extremal regions. In *Proceedings of the British Machine Vision Conference*, Vol. 1, pages 384–393.

Mikolajczyk, K. and Schmid, C. (2001). Indexing based on scale invariant interest points. In *Proceedings of the IEEE International Conference on Computer Vision*, Vol. 1, pages 525–531.

Mikolajczyk, K. and Schmid, C. (2003). A performance evaluation of local descriptors. In *Proceedings of the IEEE Computer Society Conference on Computer Vision and Pattern Recognition*, Vol. 2, pages 257–263.

Mikolajczyk, K. and Schmid, C. (2004). Scale & affine invariant interest point detectors. *International Journal of Computer Vision*, 60(1):63–86.

Mikolajczyk, K. and Schmid, C. (2005). A performance evaluation of local descriptors. *IEEE Transactions on Pattern Analysis and Machine Intelligence*, 27(10):1615–1630.

Mikolajczyk, K., Tuytelaars, T., Schmid, C., Zisserman, A., Matas, J., Schaffalitzky, F., Kadir, T., and Van Gool, L. (2005). A comparison of affine region detectors. *International Journal of Computer Vision*, 65(1–2):43–72.

Miksik, O. and Mikolajczyk, K. (2012). Evaluation of local detectors and descriptors for fast feature matching. In *Proceedings of the International Conference on Pattern Recognition*, pages 2681–2684.

Monnier, F., Vallet, B., and Soheilian, B. (2012). Trees detection from laser point clouds acquired in dense urban areas by a mobile mapping system. *ISPRS Annals of the Photogrammetry, Remote Sensing and Spatial Information Sciences*, 1(3):245–250.

Moravec, H. (1977). Towards automatic visual obstacle avoidance. In *Proceedings of the International Joint Conference on Artificial Intelligence*, Vol. 2, pages 584–584.

Nister, D. and Stewenius, H. (2008). Linear time maximally stable extremal regions. In Forsyth, D., Torr, P., and Zisserman, A., editors, *ECCV 2008, Part II*, pages 183–196. LNCS, Vol. 5303, Springer, Heidelberg, Germany.

Nixon, M. and Aguado, A. (2008). *Feature extraction & image processing (2nd edition).* Academic Press, Oxford, UK.

Novatnack, J. and Nishino, K. (2007). Scale-dependent 3D geometric features. In *Proceedings of the International Conference on Computer Vision*, pages 1–8.

Papageorgiou, C., Oren, M., and Poggio, T. (1998). A general framework for object detection. In *Proceedings of the International Conference on Computer Vision*, pages 555–562.

Prewitt, J. and Mendelsohn, M. (1966). The analysis of cell images. *Annals of the New York Academy of Sciences*, 128(3):1035–1053.

Richards, J. (2005). Analysis of remotely sensed data: the formative decades and the future. *IEEE Transactions on Geoscience and Remote Sensing*, 43(3):422–432.

Roberts, L. (1965). Machine perception of three-dimensional solids. In Tippett, J., Berkowitz, D., Clapp, L., Koester, C., and Vanderburgh, A., editors, *Optical and Electro-Optical Information Processing*, pages 159–197. MIT Press, Cambridge, UK.

Rosten, E. and Drummond, T. (2005). Fusing points and lines for high performance tracking. In *Proceedings of the IEEE International Conference on Computer Vision*, Vol. 2, pages 1508–1515.

Rublee, E., Rabaud, V., Konolige, K., and Bradski, G. (2011). ORB: an efficient alternative to SIFT or SURF. In *Proceedings of the IEEE International Conference on Computer Vision*, pages 2564–2571.

Rusu, R., Blodow, N., and Beetz, M. (2009). Fast point feature histograms (FPFH) for 3D registration. In *Proceedings of the IEEE International Conference on Robotics and Automation*, pages 3212–3217.

Rusu, R., Blodow, N., Marton, Z., and Beetz, M. (2008). Aligning point cloud views using persistent feature histograms. In *Proceedings of the IEEE/RSJ International Conference on Intelligent Robots and Systems*, pages 3384–3391.

Salti, S., Tombari, F., and Di Stefano, L. (2011). A performance evaluation of 3D keypoint detectors. In *Proceedings of the International Conference on 3D Imaging, Modeling, Processing, Visualization and Transmission*, pages 236–243.

Schmid, C., Mohr, R., and Bauckhage, C. (2000). Evaluation of interest point detectors. *International Journal of Computer Vision*, 37(2):151–172.

Schmidt, A., Kraft, M., and Kasinski, A. (2010). An evaluation of image feature detectors and descriptors for robot navigation. In Bolc, L., Tadeusiewicz, R., Chmielewski, L., and Wojciechowski, K., editors, *ICCVG 2010, Part II*, pages 251–259. LNCS, Vol. 6375, Springer, Heidelberg, Germany.

Scovanner, P., Ali, S., and Shah, M. (2007). A 3-dimensional SIFT descriptor and its application to action recognition. In *Proceedings of the International Conference on Multimedia*, pages 357–360.

Sedgewick, R. (1988). *Algorithms* (2nd edition). Addison-Wesley, Boston, USA.

Sheng, Y. and Shen, L. (1994). Orthogonal Fourier–Mellin moments for invariant pattern recognition. *Journal of the Optical Society of America A*, 11(6):1748–1757.

Sipiran, I. and Bustos, B. (2010). A robust 3D interest points detector based on Harris operator. In *Proceedings of the Eurographics Workshop on 3D Object Retrieval*, pages 7–14.

Sipiran, I. and Bustos, B. (2011). Harris 3D: a robust extension of the Harris operator for interest point detection on 3D meshes. *The Visual Computer*, 27(11):963–976.

Sobel, I. (1970). *Camera models and machine perception*. PhD thesis, Stanford University, Stanford, USA.

Steder, B., Grisetti, G., and Burgard, W. (2010). Robust place recognition for 3D range data based on point features. In *Proceedings of the IEEE International Conference on Robotics and Automation*, pages 1400–1405.

Steder, B., Grisetti, G., Van Loock, M., and Burgard, W. (2009). Robust online model-based object detection from range images. In *Proceedings of the IEEE/RSJ International Conference on Intelligent Robots and Systems*, pages 4739–4744.

Steder, B., Rusu, R., Konolige, K., and Burgard, W. (2011). Point feature extraction on 3D range scans taking into account object boundaries. In *Proceedings of the IEEE International Conference on Robotics and Automation*, pages 2601–2608.

Stöttinger, S., Goras, B., Pöntiz, T., Hanbury, A., Sebe, N., and Gevers, T. (2011). Systematic evaluation of spatio-temporal features on comparative video challenges. In Koch, R. and Huang, F., editors, *ACCV 2010 Workshops, Part I*, pages 349–358. LNCS, Vol. 6468, Springer, Heidelberg, Germany.

Stückler, J. and Behnke, S. (2011). Interest point detection in depth images through scale-space surface analysis. In *Proceedings of the IEEE International Conference on Robotics and Automation*, pages 3568–3574.

Teague, M. (1980). Image analysis via the general theory of moments. *Journal of the Optical Society of America*, 70(8):920–930.

Teh, C.-H. and Chin, R. (1988). On image analysis by the methods of moments. *IEEE Transactions on Pattern Analysis and Machine Intelligence*, 10(4):496–513.

Tola, E., Lepetit, V., and Fua, P. (2008). A fast local descriptor for dense matching. In *Proceedings of the IEEE Computer Society Conference on Computer Vision and Pattern Recognition*, pages 1–8.

Tola, E., Lepetit, V., and Fua, P. (2010). DAISY: an efficient dense descriptor applied to wide-baseline stereo. *IEEE Transactions on Pattern Analysis and Machine Intelligence*, 32(5):815–830.

Toshev, A., Mordohai, P., and Taskar, B. (2010). Detecting and parsing architecture at city scale from range data. In *Proceedings of the IEEE Computer Society Conference on Computer Vision and Pattern Recognition*, pages 398–405.

Trulls, E., Sanfeliu, A., and Moreno-Noguer, F. (2012). Spatiotemporal descriptor for wide-baseline stereo reconstruction of non-rigid and ambiguous scenes. In Fitzgibbon, A., Lazebnik, S., Perona, P., Sato, Y., and Schmid, C., editors, *ECCV 2012, Part III*, pages 441–454. LNCS, Vol. 7574, Springer, Heidelberg, Germany.

Turner, M. (1986). Texture discrimination by Gabor functions. *Biological Cybernetics*, 55:71–82.

Tuytelaars, T. and Mikolajczyk, K. (2008). Local invariant feature detectors: a survey. *Foundations and Trends in Computer Graphics and Vision*, 3(3):177–280.

Van Gool, L., Dewaele, P., and Oosterlinck, A. (1985). Texture analysis Anno 1983. *Computer Vision, Graphics, and Image Processing*, 29(3):336–357.

Viola, P. and Jones, M. (2001). Rapid object detection using a boosted cascade of simple features. In *Proceedings of the IEEE Computer Society Conference on Computer Vision and Pattern Recognition*, volume 1, pages 511–518.

Wang, H., Ullah, M., Kläser, A., Laptev, I., and Schmid, C. (2009). Evaluation of local spatio-temporal features for action recognition. In *Proceedings of the British Machine Vision Conference*, pages 127–138.

Wang, Z. and Brenner, C. (2008). Point based registration of terrestrial laser data using intensity and geometry features. *The International Archives of the Photogrammetry, Remote Sensing and Spatial Information Sciences*, 37(B5):583–589.

Weidner, U. (2005). Remote sensing systems – An overview focussing on environmental applications. In *Proceedings of the EnviroInfo 2005 Workshop on Tools for Emergencies and Disaster Management*.

Weinmann, M. (2013). Visual features: from early concepts to modern computer vision. In Farinella, G., Battiato, S., and Cipolla, R., editors, *Advanced Topics in Computer Vision*, pages 1–34. Springer, London, UK.

Weinmann, M., Hoegner, L., Leitloff, J., Stilla, U., Hinz, S., and Jutzi, B. (2012). Fusing passive and active sensed images to gain infrared-textured 3D models. *The International Archives of the Photogrammetry, Remote Sensing and Spatial Information Sciences*, 39(B1):71–76.

Weinmann, M., Jutzi, B., and Mallet, C. (2013). Feature relevance assessment for the semantic interpretation of 3D point cloud data. *ISPRS Annals of the Photogrammetry, Remote Sensing and Spatial Information Sciences*, 2(5/W2):313–318.

Weinmann, M., Weinmann, M., Hinz, S., and Jutzi, B. (2011). Fast and automatic image-based registration of TLS data. *ISPRS Journal of Photogrammetry and Remote Sensing*, 66(6):S62–S70.

West, K., Webb, B., Lersch, J., Pothier, S., Triscari, J., and Iverson, A. (2004). Context-driven automated target detection in 3-D data. *Proceedings of SPIE*, 5426:133–143.

Willems, G., Tuytelaars, T., and Van Gool, L. (2008). An efficient dense and scale-invariant spatio-temporal interest point detector. In Forsyth, D., Torr, P., and Zisserman, A., editors, *ECCV 2008, Part II*, pages 650–663. LNCS, Vol. 5303, Springer, Heidelberg, Germany.

Yu, T.-H., Kim, T.-K., and Cipolla, R. (2010). Real-time action recognition by spatiotemporal semantic and structural forests. In *Proceedings of the British Machine Vision Conference*, pages 52.1–52.12.

Yu, T.-H., Woodford, O., and Cipolla, R. (2011). An evaluation of volumetric interest points. In *Proceedings of the International Conference on 3D Imaging, Modeling, Processing, Visualization and Transmission*, pages 282–289.

Zaharescu, A., Boyer, E., Varanasi, K., and Horaud, R. (2009). Surface feature detection and description with applications to mesh matching. In *Proceedings of the IEEE Computer Society Conference on Computer Vision and Pattern Recognition*, pages 373–380.

Zhang, D. and Lu, G. (2004). Review of shape representation and description techniques. *Pattern Recognition*, 37(1):1–19.

Zhang, J. and Tan, T. (2002). Brief review of invariant texture analysis methods. *Pattern Recognition*, 35(3):735–747.

similar training example. For this purpose, a distance function has to be defined that may be based on the Euclidean distance, a general Minkowski metric or other distance metrics. Additional information is involved when applying a k nearest neighbour (k-NN) classifier (Cover and Hart, 1967), which, for each feature vector $\mathbf{x}_i$, selects the k nearest (i.e. most similar) samples in the training data and classifies according to the majority vote of their class labels. Since the induction or generalisation process is delayed to the classification of unseen feature vectors, instance-based learning techniques are also referred to as lazy learning techniques (Lopez de Mantaras and Armengol, 1998).

4.3.2 Probabilistic Learning

Probabilistic learning techniques focus on deriving an explicit underlying probabilistic model and inferring the most probable class label for each observed feature vector. Assuming random variables X_j with $j=1,\ldots,d$ and L for the d corresponding attributes x_j of a feature vector $\mathbf{x}$ and the class label l respectively, the posterior $P(L=l \mid X_1=x_1,\ldots,X_d=x_d)$ equals the product of prior $P(L=l)$ and likelihood $P(X_1=x_1,\ldots, X_d=x_d \mid L=l)$ divided by evidence $P(X_1=x_1,\ldots,X_d=x_d)$ according to Bayes' theorem:

$$P(L=l \mid X_1=x_1,\ldots,X_d=x_d) = \frac{P(L=l)\ P(X_1=x_1,\ldots,X_d=x_d \mid L=l)}{P(X_1=x_1,\ldots,X_d=x_d)} \tag{4.2}$$

Exploiting the naive assumption of all features being conditionally independent, the joint probability

$$P(L=l, X_1=x_1,\ldots,X_d=x_d) = P(L=l)\ P(X_1=x_1,\ldots,X_d=x_d \mid L=l) \tag{4.3}$$

can be simplified to

$$P(L=l, X_1=x_1,\ldots,X_d=x_d) = P(L=l)\prod_{j=1}^{d} P(X_j=x_j \mid L=l) \tag{4.4}$$

which is the general idea of a naive Bayesian (NB) classifier (John and Langley, 1995). In the classification process, an unseen feature vector is thus assigned the most probable class label according to

$$f: \hat{l} = \underset{l\in\{1,\ldots,N_C\}}{\operatorname{argmax}} P(L=l)\prod_{j=1}^{d} P(X_j=x_j \mid L=l) \tag{4.5}$$

where the probabilities $P(L=l)$ and $P(X_j=x_j|L=l)$ are determined in the training phase. However, because conditional independence is assumed, a possible

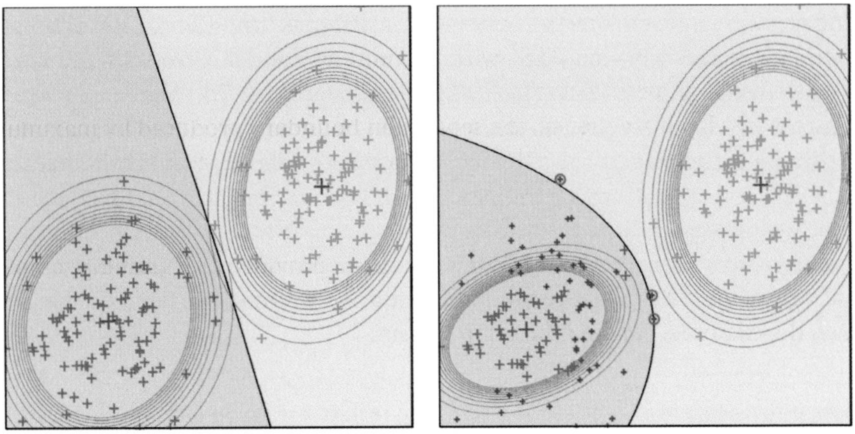

(a) (b)

Figure 4.4 The Gaussian data set from Figure 4.3a classified with a probabilistic technique (maximum likelihood classification). (a) Appropriate and thorough sampling leads to an appropriate separation. (b) Failure to represent some parts of the distribution by training data results in poor parameterisation and causes errors.

correlation between features (i.e. single entries in the feature vector) cannot be modelled appropriately.

In Figure 4.4, the Gaussian data set presented in Figure 4.3a has been classified with one of the most well-known probabilistic classifiers, the maximum likelihood (ML) classifier, for which mathematical foundations – taught in even the most basic statistical courses nowadays – are omitted here. In Figure 4.4a the distribution of both classes has been appropriately sampled, so that training data are available throughout the entire Gaussian. Therefore, means and standard deviations, required for maximum likelihood estimation, are adequately estimated and an appropriate separation boundary is derived that classifies the entire data set correctly. In Figure 4.4b, the distribution is not appropriately sampled. Only the bottom left and top right part of class 1 are represented by sampling data, missing other parts of the distribution. Accordingly, mean and standard deviations estimated from the training data deviate strongly from the true values of the entire class. Note that similar situations can occur quite frequently when collecting sample data. Consider a situation where a rather bright oak forest is defined as class 2 and has to be separated from a dark coniferous forest defined as class 1. The remote sensing scientist collects training data, labelling his position with a hand-held GPS device while wandering through the two types of forests in terrain. Despite his honest ambition to represent both classes appropriately, he cannot assess a large part of the dark coniferous forest due to the occurrence of dense raspberry shrubs below the trees. The raspberry shrubs have two effects. First, they brighten up the signature of class 1 considerably because their foliage is light green (in contrast to the dark green pine needles). Second, due to their countless sharp thorns, they do not allow our scientist to enter with his GPS[1] without running the risk of bleeding to death. The resulting pattern would be exactly a class sampling like the one shown in Figure 4.4b[2]. In consequence, the separation boundary produced by maximum likelihood classification is no longer appropriate, introducing errors, because especially bright parts of class 1 (the areas with the raspberry shrubs) are dissected from class 1 and assigned to class 2. Thus, it can be concluded that, for probabilistic learning techniques, it is crucial to provide many training samples that represent the *entire* distribution appropriately, which, as has become clear from the example, can be difficult in terrain.

4.3.3 Kernel-based Learning

Probabilistic learning techniques based on Bayes' theorem are the origin of image classification techniques. They are successfully applied even today. However, as the example in Figure 4.4b has shown, probabilistic learning can easily fail, even if all classes under consideration are perfectly Gaussian and do not overlap at all. However, in real remote sensing image classification applications, the assumption of simple distributions – like Gaussianness – is violated. Classes are frequently characterised by more complex distributions, like mixtures of Gaussians (MOG), or even more irregular distributions. Hence, it is necessary to have classification frameworks at hand that are able to distinguish arbitrarily complex distributions of data.

Therefore, during recent decades, a new paradigm of machine learning has evolved: kernel-based learning. Kernel-based learning aims to project data into a higher-dimensional feature space, where even complex and non-Gaussian data like the Ripley data set in Figure 4.3b can be more easily separated. Many kernel-based learning techniques, like the famous support vector machine (SVM), furthermore aim to use only a small subset of the training data to establish appropriate decision boundaries (Cristianini and Shawe-Taylor, 2000).

4.3.3.1 Basics of Kernel Transformations

Kernel-based learning techniques are increasingly being applied in remote sensing. They are especially purposeful, because remote sensing classification tasks frequently feature several, or all, of the following requirements:

- A high number of features is available and directly provided to the classifier.
- The number of available training data is relatively low, rendering probability density estimations a problem.
- A high number of classes, often with some of them being spectrally similar to one another, have to be distinguished.
- The classification model has to be established on few training instances in order to be computationally efficient.

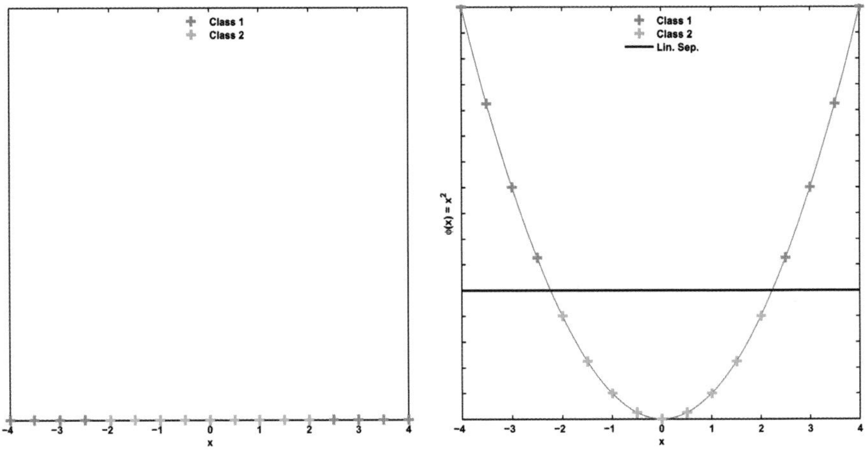

(a) (b)

Figure 4.5 Improving class separability. (a) The two classes are not linearly separable within the one-dimensional feature space. (b) The two classes become separable after introducing a non-linear mapping and thus inducing a two-dimensional feature space.

Kernel-based techniques aim to meet these requirements by projecting data from the original feature space into high-dimensional feature spaces. Assume two data points $\mathbf{x}_i, \mathbf{x}_j$ from a data set $\mathcal{X}$, in which two classes exist and are marked with labels $l_i, l_j \in [-1, +1]$. The data set is from an input feature space $\mathbb{R}^{d_1}$. The main principle of all kernel-based techniques is to project data from the original feature space $\mathbb{R}^{d_1}$ into a mathematical feature space of higher dimensionality $\mathbb{R}^{d_2}$, $d_2 \gg d_1$. For this purpose, a non-linear mapping of the type $\phi\colon \mathbb{R}^{d_1} \rightarrow \mathbb{R}^{d_2}$ is used. Within the higher-dimensional feature space $\mathbb{R}^{d_2}$, it is then easier to find a linear separation f to assign each $\mathbf{x}_i$ a label l_i, and thus perform a mapping $f\colon \mathbf{x}_i \rightarrow l_i$. Thus, by projecting data into higher-dimensional feature spaces, separability of the data is increased. Reprojecting back to $\mathbb{R}^{d_1}$, however, f is no longer linear (Camps-Valls and Bruzzone, 2005). Figure 4.5 visualises how separability can be increased by introducing non-linear transformations. In Figure 4.5a, both classes are not linearly separable within the one-dimensional feature space. After introducing the non-linear mapping $\phi(\mathbf{x}_i) = \mathbf{x}_i^2$, and thus inducing a two-dimensional feature space, the data become linearly separable[3]. Hence, kernel-based techniques generally follow the idea of introducing non-linear projections $\phi(\mathbf{x}_i)$.

Such projections follow one of the principal tenets of statistical learning theory: accuracy of classification should be increased by using a classifier of low complexity in a feature space of higher dimensionality instead of using a classifier of high complexity within a feature space of low dimensionality (Bousquet et al., 2004; Vapnik, 1999). However, kernel-based techniques differ with respect to the separation function f employed. It is important to notice that the non-linear functions $\phi(\mathbf{x}_i)$ are usually not computed explicitly. Doing so would have disadvantages due to the high-dimensionality d_2, which would increase the computational burden. Instead, the functions $\phi(\mathbf{x}_i)$ are replaced by so-called kernel functions.

$$K^{\delta}(\mathbf{x}_i, \mathbf{x}_j) = \phi(\mathbf{x}_i) \cdot \phi(\mathbf{x}_j) \tag{4.6}$$

Kernels operate within the input feature space $\mathbb{R}^{d_1}$; however, they represent scalar products within the high-dimensional feature space $\mathbb{R}^{d_2}$. Thus, kernels $K(\mathbf{x}_i, \mathbf{x}_j)$ implicitly induce spaces of the type $\mathbb{R}^{d_2}$, which are called reproducing kernel Hilbert spaces (RKHS) (Aronszajn, 1950; Wahba, 2002). This implicit mapping is sometimes referred to as kernel trick (Schölkopf, 2001).

The following explanation outlines how spaces of the type $\mathbb{R}^{d_2}$ are induced implicitly. Consider two data vectors $\mathbf{x}_i, \mathbf{x}_j$, each with two features: $\mathbf{x}_i = [x_{i1}, x_{i2}]$ and $\mathbf{x}_j = [x_{j1}, x_{j2}]$. Now, consider, that a non-linear mapping ϕ is sought, which maps both feature vectors into a feature space $\mathbb{R}^{d_2}$, where the following features exist: $[x_1^2, \sqrt{2}x_1x_2, x_2^2]$.

$$K(\mathbf{x}_i, \mathbf{x}_j) = \phi(\mathbf{x}_i) \cdot \phi(\mathbf{x}_j) \tag{4.7}$$

$$K(\mathbf{x}_i, \mathbf{x}_j) = (x_{i1}^2, \sqrt{2}x_{i1}x_{i2}, x_{i2}^2)^{\top}(x_{j1}^2, \sqrt{2}x_{j1}x_{j2}, x_{j2}^2) \tag{4.8}$$

$$K(\mathbf{x}_i, \mathbf{x}_j) = x_{i1}^2x_{j1}^2 + 2x_{i1}x_{i2}x_{j1}x_{j2} + x_{i2}^2x_{j2}^2 \tag{4.9}$$

$$K(\mathbf{x}_i, \mathbf{x}_j) = (x_{i1}x_{j1} + x_{i2}x_{j2})^2 \tag{4.10}$$

$$K(\mathbf{x}_i, \mathbf{x}_j) = (\mathbf{x}_i \cdot \mathbf{x}_j)^2 \tag{4.11}$$

Thus, by embedding into scalar products, the non-linear projection $\phi(x) = [x_1^2, \sqrt{2}x_1x_2, x_2^2]$, can be derived from $K(\mathbf{x}_i, \mathbf{x}_j) = (\mathbf{x}_i \cdot \mathbf{x}_j)^2$. The latter equation simply represents a polynomial kernel $K^{\delta}(\mathbf{x}_i, \mathbf{x}_j) = (\mathbf{x}_i \cdot \mathbf{x}_j)^{\delta}$ with $\delta = 2$, and thus a polynomial kernel of second degree. Although the polynomial kernel works on the original data $\mathbf{x}_i, \mathbf{x}_j$, each feature of $\phi(x)$ is exploited. Thus, an implicit projection from $\mathbb{R}^2 \to \mathbb{R}^3$ is realised (Ivanciuc, 2007). Note that most kernels depend on at least one parameter, and some depend on several parameters. Here, δ is a parameter of the polynomial kernel, the degree of the polynomial. Later in this chapter, note that δ will be used to denote kernel parameters in general. Such kernel parameters have to be defined by the human operator. Not every function can be used as a kernel function, though. The constraints given for a mathematical function to serve as a scalar product $\phi(\mathbf{x}_i) \cdot \phi(\mathbf{x}_j)$ within an RKHS are given by the theorem of Mercer (1909). Mercer's theorem states that the functions serving as kernel functions need to be symmetrical and positive semi-definite:

$$\text{symmetrical:} \quad K(\mathbf{x}_i, \mathbf{x}_j) = K(\mathbf{x}_j, \mathbf{x}_i), \quad \forall \mathbf{x}_i, \mathbf{x}_j \tag{4.12}$$

$$\text{positive semi-definite:} \quad \sum_{i=1}^{N}\sum_{j=1}^{N} K(\mathbf{x}_i, \mathbf{x}_j)c_i c_j \geq 0, \quad \forall \mathbf{x}_i, \mathbf{x}_j, \quad \forall c_i, c_j \in \mathbb{R} \tag{4.13}$$

If a function is symmetrical and positive semi-definite, then the kernel[4] serves as a scalar product of the type $\phi(\mathbf{x}_i) \cdot \phi(\mathbf{x}_j)$ within RKHS $\mathbb{R}^{d_2}$.

It can be concluded that kernel-based learning aims to use higher-dimensional feature spaces to overcome the limitations of non-linearity in the input feature space that has been shown to produce errors in some of the examples of this chapter. For instance, see Figure 4.3b. It quickly becomes clear that approaches aiming to introduce simple separation criteria, like nearest neighbour, Gaussian or even MOG approaches, will become infeasible due to the complexity of the data set.

4.3.3.2 Basics of Kernel-based Learning

A large number of kernel-based classification techniques have been proposed in the literature. Within this chapter, the well-known support vector machine (SVM) is considered in detail. However, many of its extensions and improvements, like the import vector machine (IVM) (Zhu and Hastie, 2005) or the relevance vector machine (RVM) (Tipping, 2000), follow similar basics. Therefore, instead of starting with an explanation of the mathematical foundations of SVM, a more general framework, valid for SVMs, IVMs and RVMs and described by Tipping (2001), is presented first.

The representer theorem of statistical learning theory – which is an important fundamental of SVM and similar techniques – shows that minimisation of the empirical risk function $f(\mathbf{x}_k, \omega)$, given in some RKHS, can be formalised as a finite linear combination of kernel values of $\mathbf{x}_k$ with training data $\mathbf{x}_i$ (Kimeldorf and Wahba, 1970). Thus, methods like SVM, IVM and RVM define generalised linear functions of the following type to classify the data:

$$l_k = f(\mathbf{x}_k; \omega) = \sum_{i=1}^{N} \omega_i \phi(\mathbf{x}_k) + \omega_0 \tag{4.14}$$

Here, ω represents a vector of linear weights ω_i. Hence, such models are called linear in the parameters (Tipping, 2001). ω_i are sought to be $\omega_i = 0$ for most of the training data. Furthermore, as stated above, $\phi(\mathbf{x}_k)$ are non-linear basis functions, represented by kernel function $\phi(\mathbf{x}_k) = K(\mathbf{x}_i, \mathbf{x}_k)$. ω_0 is a bias parameter for regularisation, needed by some techniques like SVM, although not for others like IVM or RVM, for which $\omega_0 = 0$. Since the basis functions $\phi(\mathbf{x}_k) = K(\mathbf{x}_i, \mathbf{x}_k)$ are computed even before training, parameterising the model merely consists in defining suitable parameters for ω_i. Each kernel-based technique then applies a classification rule on Equation 4.14. Another common point of many techniques is that they work without a rejection class in the first place. Thus, each data point is classified and no data points are left without labels. However, a rejection class is easy to define. For instance, a lower threshold on posterior probabilities $P(l_k|\mathbf{x}_k)$ can be defined, below which the data point is left unlabelled. These circumstances are important to many remote sensing developers aiming to work with a meaningful rejection class.

4.3.3.3 Introduction to the SVM

Based on these general explications of kernels and kernel-based learning, the most frequently used technique of kernel-based learning shall be exemplified: the SVM proposed by Vladimir Vapnik and his colleagues (Boser et al., 1992; Cortes and Vapnik, 1995), which has become one of the standard classification techniques of contemporary remote sensing (Mountrakis et al., 2011; Plaza et al., 2009). Although SVM is based on relatively few key ideas, the complete mathematical foundations of statistical learning theory (Vapnik, 1999) and the theory of Hilbert spaces (Berlinet and Thomas-Agnan, 2004) are extensive and it is not necessary to understand them for the application chapters of this volume. Therefore, this section focuses on the most important foundations of SVM. SVM was later reformulated by Schölkopf et al. (2000). However, herein, the standard formulation of SVM shall be explained (Boser et al., 1992; Cortes and Vapnik, 1995).

4.3.3.4 A Non-mathematical Example

Before providing the mathematical foundations, an example of SVM classification is given in Figure 4.6 to show what SVM is basically about. Figure 4.6a shows the maximum likelihood classification from Figure 4.4 for comparison. Figure 4.6b shows a linear SVM solution. Instead of using the distribution (estimating parameters such as mean and standard deviation), SVM uses just a few training samples from each class. These are the training samples closest to the respective other class and are called support vectors (SVs). Between these support vectors, a decision boundary is established, as it is assumed that the real frontier between the classes must be lying somewhere between the most "borderline points".

4.3.3.5 Hard Margin Linear SVM

As has been shown, only SVs are used for separation, and all the other training data are not required at all. The separation boundary is established in a way that a maximum margin is left free of training data for reasons of generalisation. Assuming that some data, for example belonging to class 1, may be less typical for class 1 than its training data (and thus even closer to class 2), a maximum data-free space between both classes ensures that even less typical data will be classified appropriately. The SVM is given as a primal and dual formulation. Its primal formulation is

$$\min_{\phi_0, w} \frac{||w||^2}{2} \tag{4.15}$$

$$l_i(\phi_0 + w \cdot \mathbf{x}_i) \geq 1 \tag{4.16}$$

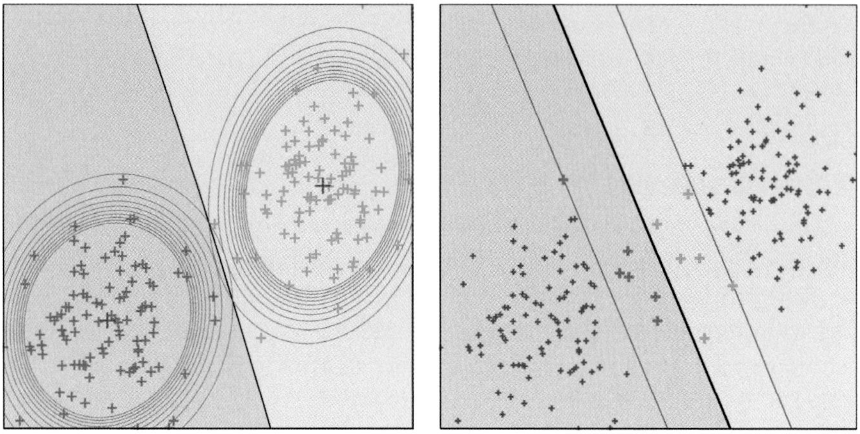

(a) (b)

Figure 4.6 Comparison between (a) probabilistic classification and (b) kernel-based classification. This kernel-based approach (an SVM) uses only a few points to establish a decision boundary, without aiming to estimate statistical distributions.

Here, w is a normal vector that is orthogonal to the separating plane between the two classes in $\mathbf{x}_i$ labelled by $l_i \in [-1, +1]$. It can be shown that maximising the width of the margin equals minimising the vector norm of w (i.e. minimising Equation 4.16). ϕ_0 is a bias, which shifts the separating plane from the origin of the feature space to the area between the two classes. Besides this, note that ϕ_0 is one of the free parameters of the generalised linear model in Equation 4.14.

The classification rule for the case of data, that are linearly separable, is $\text{sign}(\phi_0 + w \cdot \mathbf{x}_i)$. Via the signum function, the output of $f(\mathbf{x}_i) = \phi_0 + w \cdot \mathbf{x}_i$ is mapped to $[-1, +1]$. Hence, so far, each of the training data belonging to the $+1$ class has to be located on the positive side of the separating plane:

$$\text{sign}(f(\mathbf{x}_i)) = +1, \forall i : l_i = +1 \tag{4.17}$$

By analogy, for each of the training data belonging to the -1 class,

$$\text{sign}(f(\mathbf{x}_i)) = -1, \forall i : l_i = -1 \tag{4.18}$$

must hold true. Thus, for the entire training data, the equation:

$$\text{sign}(f(\mathbf{x}_i)) = l_i \tag{4.19}$$

must hold true. This claim leads to the constraint given in Equation 4.16, and is frequently referred to as a "hard margin approach".

4.3.3.6 Soft Margin Linear SVM

In the hard margin approach, the SVM aims to maximise a data-free space *and* correctly classify all the training data. Note that the two objectives can conflict.

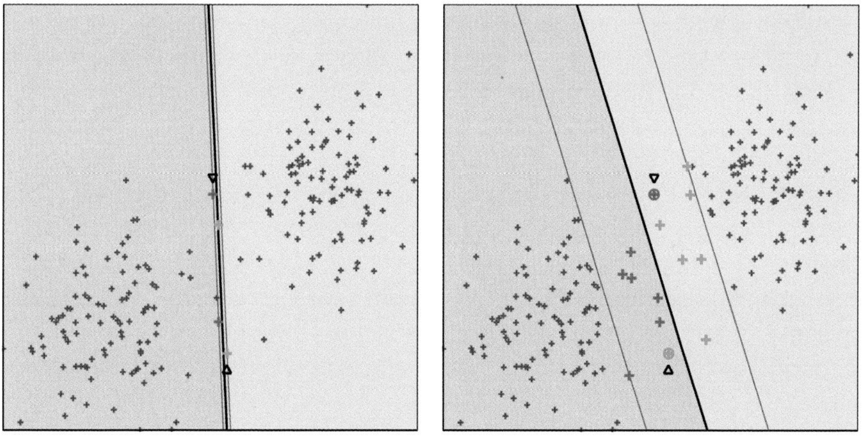

(a) (b)

Figure 4.7 Hard margin linear SVM and soft margin linear SVM. (a) When aiming to correctly classify all the training data, the margin becomes very narrow. (b) When aiming to produce a model that accepts few errors and thus allows a larger margin, the capability to generalise is improved.

In Figure 4.7, one point has been added to each of the classes from Figure 4.3a. These points lie very close to the respective other class; for example, because of struggling to escape from the raspberry shrubs in which he was stuck, the scientist from our aforementioned example has some false labels (i.e. a point belonging to class 1 is labelled with class 2 and vice versa). In Figure 4.7a, although it is still feasible to correctly classify each training data point, the margin becomes very narrow when trying to assign the two new points to their (falsely assigned) class labels. Instead of narrowing the margin and losing generalisation, it may be more appropriate to produce a model that accepts errors on the two points and – by doing so – allows a larger margin, as shown in Figure 4.7b. Such an approach is called a "soft margin approach", and its foundations are given in the following. One possibility of defining such a soft margin SVM is the cost-SVM by Boser et al. (1992) and Cortes and Vapnik (1995). The cost-SVM is also given as a primal and dual formulation. Its primal formulation is

$$\min_{\phi_0, w} \frac{||w||^2}{2} + C \sum_{i=1}^{N} \xi_i \tag{4.20}$$

$$l_i(\phi_0 + w \cdot \mathbf{x}_i) \geq 1 - \xi_i \tag{4.21}$$

In many cases, training data are either affected by noise, or have even been falsely labelled. Thus, the aforementioned claim to have each data point of the training data on the "correct side" of the plane may be hard to realise or render the separating plane unnecessarily complex. Therefore, in many cases, it is helpful to modify the aforementioned constraint. Instead of employing the constraint in its strict form, it is relaxed by allowing some training data to be on the "incorrect side" of the separating plane. Hence, some errors on the training data are allowed in favour of generalisation. The modified constraint, given in Equation 4.21, thus requires $l_i f(\mathbf{x}_i)$ only to be larger than $1 - \xi_i$. ξ_i are semi-positive variables, called slack variables. If a training point is on the correct side of the separating plane $(f(\mathbf{x}_i)) = l_i$ holds true and $\xi_i = 0$. Otherwise $\xi_i > 0$, where ξ_i is simply the distance of point i to the "correct side" of the separating plane[5].

In order to easily integrate the soft margin approach into optimisation, all ξ_i are summed and weighted using the cost factor C, which has to be set by the operator. The higher C is, the stronger is the dependency of optimisation on a correct classification of the majority of the training data. The smaller C is, the more training errors are allowed in favour of generalisation when solving the optimisation problem.

4.3.3.7 Parameterisation by Cross Validation

Since C – just like the kernel parameter(s) δ – is not optimised when solving for the generalised linear model in Equation 4.14, SVM training has to be repeated several times over the set of combinations of $[C, \delta]$. This is usually done by applying cross validation, or a similar heuristic, tuning an optimal combination of $[C, \delta]$ for some predetermined accuracy value (frequently, overall accuracy or the κ coefficient).

4.3.3.8 Soft Margin Non-linear SVM

So far, the primal formulation of SVM has been discussed. However, to more thoroughly comprehend the property of the SVM being dependent only on a few training data, and to make the step to non-linearity by kernels, it is helpful to focus also on the dual formulation[6]. A full derivation of the dual formulation from the primal formulation, using Lagrangian methods, can be found in (Burges, 1998) or (Ivanciuc, 2007) and is not included herein.

Generally, the dual formulation is derived by integrating the constraints of Equation 4.21 directly into the optimisation problem by using Lagrange multipliers. Usually, but not necessarily (Chapelle, 2007), the dual formulation is used to solve the SVM optimisation problem (Bottou and Lin, 2007). The dual formulation of the cost-SVM is

$$\max_{\omega_0, w} \sum_{i=1}^{N} \lambda_i - \frac{1}{2} \sum_{i=1}^{N} \sum_{j=1}^{N} \lambda_i \lambda_j l_i l_j \mathbf{x}_i \mathbf{x}_j \tag{4.22}$$

$$0 \leq \lambda_i \leq C, \forall i \sum_{i=1}^{N} \lambda_i l_i = 0 \tag{4.23}$$

Because all other variables are fixed, only the Lagrange multiplier λ_i of each training data point $(\mathbf{x}_i, l_i)$ is defined as explained before. For the majority of training data points, $\lambda_i = 0$ holds true. Thus, the right-hand side of Equation 4.22 becomes

$$\lambda_i \lambda_j l_i l_j \mathbf{x}_i \mathbf{x}_j = 0, \forall i, j : \lambda_i = 0 \vee \lambda_j = 0 \quad (4.24)$$

so that only zero is added to the double sum of the equation. Hence, all training data points with $\lambda_i = 0$ do not influence the optimisation problem at all. However, for a small number of training data points, $0 < \lambda_i \leq C$ holds true. These training data points are exactly the SVs that have been mentioned before, but which may be more fully understood only now. SVs only influence the optimisation problem when multiplied with other SVs. For SVs, the second constraint $\Sigma \lambda_i l_i = 0$ must hold, since otherwise, the optimisation biases towards either class $+1$ or class -1. A large number of algorithms for solving the dual formulation have been proposed (Bottou and Lin, 2007), with the sequential minimal optimisation (SMO) of Platt (1998) being one of the most commonly used algorithms. Hence, a good SVM solution only depends on relatively few data points in the training data set (the SVs). Besides this fact, using the dual formulation, the projection from the input feature space $\mathbb{R}^{d_1}$ to the higher-dimensional feature space $\mathbb{R}^{d_2}$ using $\phi : \mathbb{R}^{d_1} \to \mathbb{R}^{d_2}$ becomes straightforward. In the dual formulation, feature vectors $\mathbf{x}_i$ and $\mathbf{x}_j$ appear as scalar products $\mathbf{x}_i \cdot \mathbf{x}_j$. Such scalar products can be replaced by the high-dimensional scalar products $\phi(\mathbf{x}_i) \cdot \phi(\mathbf{x}_j)$. For the reasons explained above, this is done by using kernel functions $K(\mathbf{x}_i, \mathbf{x}_j) = \phi(\mathbf{x}_i) \cdot \phi(\mathbf{x}_j)$. Hence, in the dual formulation, only the scalar products of the feature vectors have to be replaced by kernel functions, not changing anything else in the optimisation problem or its constraints.

$$\max_{\omega_0, w} \sum_{i=1}^{N} \lambda_i - \frac{1}{2} \sum_{i=1}^{N} \sum_{j=1}^{N} \lambda_i \lambda_j l_i l_j K(\mathbf{x}_i, \mathbf{x}_j) \quad (4.25)$$

Then, to assign a class label l_k to a previously unlabelled data point $\mathbf{x}_k$, only the kernel function of $\mathbf{x}_k$ with each training data $\mathbf{x}_i$ has to be computed, weighted by λ_i, summed, and shifted by ϕ_0 away from the origin[7].

$$l_k = \text{sign}\left(\sum_{i=1}^{N} \lambda_i l_i K(\mathbf{x}_i, \mathbf{x}_k) + \phi_0 \right) \quad (4.26)$$

Thus, the weighted sum in Equation 4.26 represents the general model from Equation 4.14, where $\omega_i = \lambda_i l_i$. The bias parameter ω_0 is the mean of the loss function of the SVM over all SVs:

$$\omega_0 = l_j - \sum_{i=1}^{N} \lambda_i l_i K(\mathbf{x}_i, \mathbf{x}_j), \forall j : \lambda_j > 0 \quad (4.27)$$

Applying the signum function over the weighted sum leads to the classification rule of the SVM.

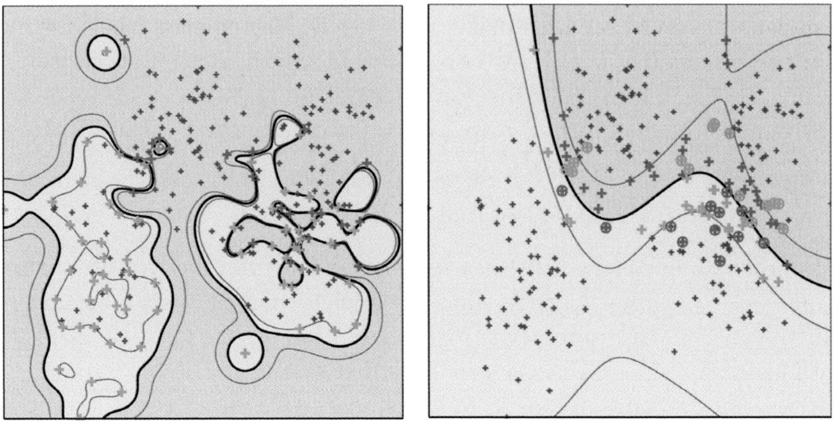

(a) (b)

Figure 4.8 Example of non-linear SVM classification of the Ripley data set. (a) Hard margin classification, correctly classifying all training data points but losing generality. (b) Soft margin classification, allowing some training errors in favour of generalisation.

At this point, we hope the main SVM principles are thoroughly understood so that it has become clear how non-linearity is introduced using the kernel function. Of course, the kernel version of the SVM, just as the linear SVM, can use the same soft margin principle to improve generalisation by allowing some errors on the training data. An example of a kernel SVM is shown in Figure 4.8. Figure 4.8a shows a highly non-linear hard margin classification of the Ripley data set from Figure 4.3. As becomes obvious, the separation boundary of the SVM can become arbitrarily complex with correct assignment of all training data points to their classes. Again, as shown in Figure 4.8b, allowing some training errors produces an SVM model that classifies most points correctly, while generalising well.

4.3.3.9 Multi-class SVM

The classification rule of the SVM, explained above, shows two immediate advantages of the dual formulation. Since, for most training data $\lambda_i = 0$, the weighted sum has to be computed using only very few points, the SVs, thus strongly reducing computational load. Furthermore, the dual formulation is feasible within explicit computation of the normal vector w, which, of course, can be derived from it if required. However, the dual formulation also highlights two preliminary disadvantages of SVM, for example, over probabilistic methods.

Via the signum function, the classification rule maps the data to $[-1, 1]$; omitting it would only change the classification rule to theoretically map to $[-\infty, +\infty]$. In both cases, no posterior probabilities $P(l_k|x)$ are produced by the SVM so far. However, for subsequent analyses, analysis of membership probabilities, applying probabilistic classification and, as such, having posterior probabilities at hand is desirable. There is a second disadvantage, though. The formulation of SVM requires that $l_i \in [-1,+1]$. Having class labels for more than two classes (i.e. $l_i \in [1,K]$) prohibits solving the optimisation problem. For instance, the constraint of the primal formulation $l_i(\omega_0+w\cdot\mathbf{x}_i) \geq 1-\xi_i$ or the constraint of the dual formulation $\Sigma\lambda_i l_i = 0$ would become infeasible. Hence, restricting to $l_i \in [-1,+1]$ defines the SVM as a binary classifier. However, in remote sensing application scenarios, more than two classes usually have to be separated.

In order to do so, the SVM has to be extended from the two-class case to the K-class case. This is usually done by applying classification cascades that split the K-class case into several two-class cases (Cheong et al., 2004; Hsu and Lin, 2002). Nonetheless, alternatives that extend the SVM to the K-class case *without* cascades are given in (Weston and Watkins, 1998), (Franc and Hlavac, 2002), and (Guermeur et al., 2000). In (Braun, 2009) and (Braun et al., 2010), several strategies have been compared and the one-against-one cascade proved to be empirically superior, as backed up by findings of other authors like Debnath et al. (2004), Melgani and Bruzzone (2004), Waske et al. (2010) and He et al. (2005). The one-against-one cascade chooses two out of K classes and trains one SVM model for these two classes. Subsequently, an SVM model is trained for each of the possible permutations of $t = K(K-1)/2$ classes. During the classification phase, each of the previously unlabelled data points $\mathbf{x}_k$ receives one label l_k^t of each of the two-class SVM models. Each of these labels is recorded in a label vector θ_k. Afterwards, each data point receives a final label l_k, which is the mode of θ_k (i.e. the label that was most frequently assigned to it). Therefore, this approach is called "max wins strategy". The cascades can also be exploited to fuse semantically different data (Braun et al., 2012).

4.3.3.10 Probabilistic Outputs

As outlined before, the $f(\mathbf{x}_k;\omega)$ values do *not* represent probabilities, but, due to their dependency on δ and C, they are theoretically scaled to $[-\infty,+\infty]$. However, after classification, having posteriors often becomes helpful. Therefore, posterior probabilities have to be derived from the SVM outputs.

For this purpose, $f(\mathbf{x}_k;\omega)$ of all two-class SVM models has to be transformed into a meaningful probability. Several approaches for doing so have been published and are summarised by Platt (1999). A rather simple approach is given by computing $P(y=1|y=i \vee y=j,f) = 1/(1+\exp(-f))$ (Wahba, 1992, 1999). In contrast, Platt (1999) fits a sigmoid function to $f(\mathbf{x}_k;\omega)$. His model $P(y=1|y=i \vee y=j,f) = 1/(1+\exp(\Lambda_1 f + \Lambda_2))$ depends on two parameters, Λ_1, Λ_2, which are determined by maximum likelihood estimation, minimising the negative logarithmic likelihood (NLL) of a training data set. Once Λ_1, Λ_2 are determined, $P(y=1|y=i \vee y=j,f)$ for each of the two-class models can be established. Currently, one of the most frequently used SVM packages in remote sensing research is LibSVM (Chang and Lin, 2011). This package modifies the approach of Platt (1999) by using a five-fold cross validation to estimate Λ_1, Λ_2 (Lin et al., 2007). From the individual $P(y=1|y=i \vee y=j,f)$

of each two class models, a posterior probability for each of the K classes $P_k(y=k|x), k=1,\ldots,K$ has to be determined. For this purpose, a pairwise coupling approach is followed (Wu et al., 2004). In order to simplify notification, we set $r_{ij}=P(y=1|y=i \vee y=j, f)$. Then, the following target function is minimised:

$$\min_{p} \frac{1}{2} \sum_{i=1}^{K} \sum_{j:j \neq i}^{K} (r_{ji} \cdot P_i - r_{ij} \cdot P_j)^2 \tag{4.28}$$

$$P_i \geq 0 \forall i, \sum_{i}^{K} P_i = 1 \tag{4.29}$$

Once this optimisation problem is solved, posterior probabilities for each class are available: $P_k(y=k|x), k=1,\ldots,K$. Note that, although these posterior probabilities are derived from the membership function $f(\mathbf{x}_k;\omega)$ of the SVM, they are established from parameter estimations derived from the training data. Due to this special approach, the membership function, defined by $\text{argmax}(P_k(y=k|x))$, is not identical to the membership function derived from $(f(\mathbf{x}_k;\omega))$ for each x. This has led some authors, like Tipping (2001), to criticise that the method proposed by Platt (1999) is not necessarily a good approximation of posterior probabilities. Such criticism has inspired a vivid debate on probabilistic enhancements of SVM and has led to the proposal of new techniques like IVM and RVM.

4.3.4 Ensemble Learning

Having focused on instance-based learning, probabilistic learning and kernel-based learning so far, we finally consider ensemble learning, which is based on the idea of strategically generating a set of weak learners and combining them to create a single strong learner. A weak learner in this context is a classifier whose predicted labels are only slightly correlated with the true labels (i.e. the labelling is only slightly better than random guessing). In contrast, a strong learner represents a classifier whose predicted labels are strongly correlated with the true labels.

Typically, the same base learner is used to generate multiple hypotheses. These hypotheses are subsequently combined to form a (hopefully) better hypothesis representing the predicted label[8]. Since evaluating multiple learners requires more computational effort than evaluating a single learner, most of the approaches for ensemble learning exploit fast algorithms such as decision trees as weak learners. For the combination of weak learners, different strategies have been proposed such as bagging (Breiman, 1996), boosting (Schapire, 1990) and their numerous variations.

4.3.4.1 Bagging

An intuitive way for combining a set of weak learners has been proposed with bootstrap aggregating, which is commonly referred to as bagging (Breiman,

1996). Using bootstrapped replica of the training data (i.e. subsets of the complete training data that are randomly drawn with replacement; Efron, 1979), diversity is obtained by training a weak learner of the same type for each subset of the training data. Because the resulting individual classifiers are likely to generate multiple hypotheses, the respective straightforward approach for combining all weak learners to a strong learner is obtained by taking the majority vote over all hypotheses.

The most popular example for bagging is represented by a random forest classifier (Breiman, 2001), which aggregates hypotheses derived via decision trees, which, in turn, are trained over different distributions of the training data. In the training phase, individual decision trees are trained on randomly selected feature subsets of the given training data. Thus, the trees are all randomly different from one another, which results in a de-correlation between individual tree predictions and thus improves generalisation and robustness (Criminisi and Shotton, 2013). For a new feature vector, each decision tree votes for a single class and a respective label is subsequently assigned according to the majority vote of all decision trees. The parameters to be specified are the number of involved decision trees, the minimum allowable number of training data points for a tree node to be split, the number of active variables to be used for the test in each node, and the maximum tree depth. As we may often face an unbalanced distribution of training examples per class in the training set, which may have a detrimental effect on the training process (Criminisi and Shotton, 2013), a class re-balancing can be introduced that consists of resampling the training data in order to obtain a uniform distribution of randomly selected training examples per class. The alternative would be to exploit the known prior class distribution of the training set for weighting the contribution of each class.

4.3.4.2 Boosting

An alternative to bagging has been presented with boosting (Schapire, 1990), which also relies on the majority vote of multiple weak learners created by resampling the data. In contrast to bagging, however, boosting is based on incrementally deriving an ensemble of classifiers over consecutive iterations and thus over different distributions of the training data. In each iteration, a subset of the complete training data is selected to train a weak learner and get a weak hypothesis with low error with respect to the true labels. Consequently, the selected distribution of the training data specifies the relative importance of each training example for the current iteration. After a certain number of iterations, the weak hypotheses are combined to a single vote as the majority vote over all hypotheses. Since boosting was originally proposed for binary classification problems, an extension to multi-class classification has been proposed with "adaptive boosting" (Freund and Schapire, 1997), which is commonly referred to as AdaBoost and widely used for many applications. Unlike bagging, where the hypotheses of weak classifiers are considered with equal weights, AdaBoost generates a weighted majority vote in which the weight of each weak hypothesis represents a function of its accuracy (Freund and Schapire, 1997).

4.4 One-class Versus Multi-class Classification

One-class classification is conceptually different from multi-class classification. Multi-class classification considers N different classes and tries to separate them *from one another.* In contrast, one-class classification considers only a single class and tries to separate it *from the remainder of the data space.* Consider a simple example: multi-class classification tries to distinguish trees, stones, water and soil, while one-class classification tries to distinguish trees from anything that is not a tree.

So far, we have concentrated on outlining the advantages of modern multi-class classifiers like cost-SVM. The question now is, why, if these methods really work that well, do we consider changing to an entirely different concept of classification?

One of the strongest arguments in favour of one-class classification comes from the realities of training data assignment. In many cases, such as land-use scenarios in remote sensing or object recognition scenarios in computer vision, one is not interested in the plethora of objects that can be found in the images. For instance, a robot only needs to select *one* certain component required for some technical process and can ignore all other components visible via its camera. Or, coming back to the aforementioned example, our plant geographer wonders why he should rupture his skin and clothes in the raspberry shrubs below the pine trees if his actual task is only to map the *oak* forests. In both examples, an image analysis framework ready to detect one kind of object *without reference* to other kinds of objects may be more appropriate. Also, consider the hundreds of image classes a 185 km × 185 km Landsat scene may contain and how frustrated one can get assigning training areas and tuning a classifier for each, if, in fact, you are only interested in determining the amount of water surfaces in the scene.

Such situations motivate the usage of one-class classification for some applicants of image analysis. For methods developers, developing one-class methods is a challenging and enriching field of work, since it may in fact be just as, or even more intricate than, multi-class classification[9].

One-class classification – sometimes also referred to as "domain description" or "absolute classification" (whereas multi-class classification is referred to as "relative classification") – is another world full of methods, algorithms and techniques that would easily fill another book. Here, only a single technique is described in analogy to multi-class classification: support vector domain description (SVDD).

For this purpose, see Figure 4.9a, where we see only one class from Ripley's data set (Figure 4.3b). In the following, we will see how SVDD can be used to separate this class from the remainder of the data space.

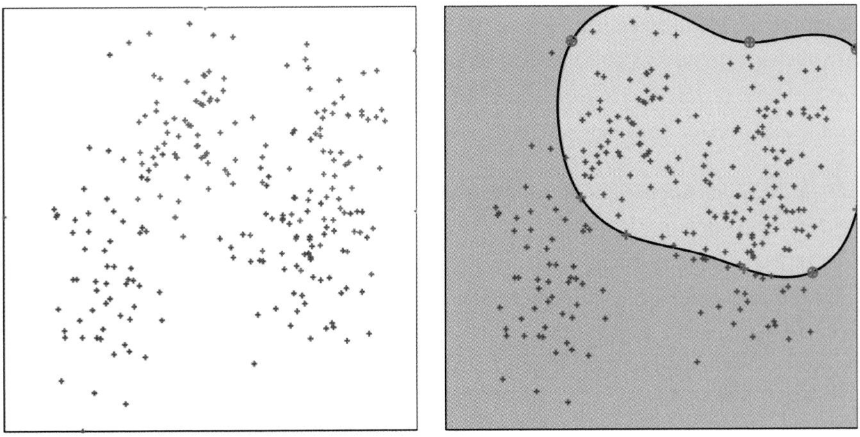

(a) (b)

Figure 4.9 Example of one-class SVM classification (SVDD) of the Ripley data set. (a) Data set with one class. (b) SVDD separation of this class from the remainder of the data space.

4.4.1 One-class SVMs

In this section, we will give a short introduction to the principal ideas behind SVM. We will focus on the foundations of the one-class SVM and its analogy to the well-established cost-SVM (Burges, 1998). The one-class SVM was proposed by Schölkopf (2001) to estimate the support of unknown distributions in high-dimensional feature spaces. The general idea behind the one-class SVM – and at the same time the most important difference to the cost-SVM – is that it is not designed to separate two classes from each other, but to precisely describe the distribution of one single class. Given a training set of $\mathbf{x}_i$ instances without any class information,

$$\mathbf{x}_i \in \mathbb{S}^\tau, \quad i = 1, \ldots, l \tag{4.30}$$

the one-class SVM constructs an enclosing hypersphere of minimum volume within a space $\mathbb{H}^\psi$ of higher dimensionality for the given set (i.e. $\psi \geq \tau$)[10]. This is accomplished by introducing a non-linear feature mapping $\Phi : \mathbb{S}^\tau \longrightarrow \mathbb{H}^\psi$. Note that just like the cost-SVM, the one-class SVM solves this task by substituting Φ with kernel functions. To find this optimal enclosing hypersphere, the origin is used as the only reference point outside the class (cf. Figure 4.9). In Equation 4.32, it can be shown that the distance from the origin to the hypersphere can be maximised by minimising the norm of vector w and the parameter $\rho \in [0, 1]$. Once w and ρ are found, the decision function for new and unlabelled points is given by

$$f(x) = \text{sign}(w \cdot \Phi(\mathbf{x}_i - \rho)) \tag{4.31}$$

Thus, the one-class SVM delivers a decision function that takes the value $+1$ in a region that includes the majority of training data and takes the value -1 for only very few data (considered non-supporting training instances). Introducing a parameter ν delivers a trade-off between the fraction of non-supporting patterns and the size of the positive region. The parameter $\nu \in (0, 1]$ serves at the same time as an upper bound on the fraction of non-supporting patterns and as a lower bound on the number of support vectors (Schölkopf, 2001):

$$\min_{w,\xi,\rho} \quad \frac{1}{2}\|w\| - \rho + \frac{l}{\nu l}\sum_{i=1}^{l} \xi_i \tag{4.32}$$

$$\text{subject to} \quad w^{\mathsf{T}}\Phi(\mathbf{x}_i) \geq \rho - \xi_i$$

$$\xi_i \geq 0, \quad i = 1, \ldots, l$$

A solution is found by solving a quadratic optimisation problem (see Equation 4.32). Usually, this problem is solved in its dual form (given in Equation 4.33), derived by the Lagrangian:

$$\min_{\alpha} \quad \frac{1}{2}\alpha^{\mathsf{T}} Q\alpha \tag{4.33}$$

$$\text{subject to} \quad 0 \leq \alpha_i \leq \frac{1}{\nu l}, \quad i = 1, \ldots, l$$

$$\min_{\alpha} \frac{1}{2}\alpha^{\mathsf{T}} Q\alpha = 1$$

where $Q_{i,j} = K(\mathbf{x}_i, \mathbf{x}_j) \equiv \Phi(\mathbf{x}_i)^{\mathsf{T}}\Phi(\mathbf{x}_j), \quad i, j \in 1, \ldots, l$ represents a kernel function that performs the transformation $\Phi(\mathbf{x}_i)$ into the higher-dimensional space implicitly (Burges, 1998). These kernels have proprietary parameters (like γ for the width of the Gaussians in the radial basis function kernel) that need to be adjusted together with ν by performing a grid search. The computation for the SVM grid search is time-consuming and lengthens drastically with the amount of training information. This explains the numerous efforts to reduce the necessary training data to a minimum. As we can see, the one-class SVM is an absolute classifier that makes use of the same mathematical foundations as the cost-SVM, but is adapted to detect only single classes employing the ν-strategy for error penalisation.

4.5 Conclusions

In this chapter, a set of state-of-the-art classification techniques used in contemporary remote sensing science have been described. Before deciding on a particular classifier, several decisions have to be taken. Two of them, namely the dichotomy between unsupervised and supervised classification, and the dichotomy between rule-based and exemplar-based learning, have been ignored by this chapter, opting for the latter option of each possibility. Besides these

visualisation and is found at http://www.youtube.com/watch?v=3liCbRZPrZA (last accessed 5 December 2014).

4. For approaches using kernels that do *not* satisfy Mercer's theorem, see (Ong et al., 2004), (Lin and Lin, 2003) and (Bahlmann et al., 2002).
5. At this point, one may figure out that the slack variables represent the weights ϕ_i in Equation 4.14. This is not exactly the case. The weights ϕ_i of the SVM are defined in the following, when the dual formulation is explained.
6. Note that switching from the primal to the dual formulation here is not strictly necessary. On the one hand, the SVM can be perfectly understood both in the primal and dual formulation. Hence, we could go on from here using the primal formulation. On the other hand, we could have also switched to the dual formulation right at the beginning. Therefore, switching from the primal to the dual formulation at this point is purely arbitrary and is done here so that readers can more easily capture the main ideas of the SVM.
7. Of course, since for most training data $\lambda_i = 0$ holds, it makes immediate sense to sum up only over the SVs.
8. A simple example again illustrates this point. Imagine you are looking for the main station in a city you have never visited before. Using regular (non-ensemble) learning is like looking for a *single person* who looks most competent about the location of buildings within the city, asking this person for the way, and confidently walking in the exact direction the person mentioned. Ensemble learning is like asking *several persons* without having a particularly high confidence in any of them and then walking in the direction that best suits all the directions they have mentioned (ignoring some unusual suggestions of a few persons that might not be trustworthy).
9. Instead of finding *differences* between objects, one-class classification – in a philosophical sense – is a task that tries to describe the *essential properties* of one class of objects. To see why this may be more intricate, consider the following simple example. Quickly write down some differences that distinguish a chair from a bed (multi-class classification). You will certainly not have a hard time finding some criteria that are sufficient to keep any person from trying to sleep on a chair (accidently, at least). Then, just try to write down the *essential properties* that *make* a chair a chair and exclude *any other* objects from being a chair (remembering all the fascinating shapes of chairs modern artists and furniture designers have come up with).
10. $\psi = \tau$ for the linear kernel $K(\mathbf{x}_i, \mathbf{x}_j) = \langle \mathbf{x}_i, \mathbf{x}_j \rangle, \quad i, j \in 1, \ldots, l$

Context-based Classification

CHAPTER 5

Franz Rottensteiner

So far we have only considered the case that all the image primitives (pixels or segments) are classified independently from each other. A typical example of the results to be expected for an independent pixel-based classification of a satellite image is shown in Figure 5.1. Obviously, there are quite a few isolated pixels that are assigned to a wrong class, and, as a consequence, the result appears to be rather noisy. The overall accuracy of the classification, that is, the percentage of pixels assigned to the correct class, is 73.2%. A strong local variation of class labels as shown in Figure 5.1 is very unlikely to occur in this type of imagery. In order to improve the results, *context* can be considered in the classification process.

According to Wolf and Bileschi (2006), context can be defined as *any information relevant to the detection task but not directly due to the physical appearance of the object.* According to Hinz and Baumgartner (2003), one can differentiate *global context* and *local context.* Global context is, for instance, related to the type of scene that is to be analysed, whereas local context considers local information. In remote sensing applications, *geometrical context,* describing the typical relative alignment of objects in a scene, is of paramount importance.

There are several options for considering context in image analysis. In a model-based scenario, the object model can be expanded by *context objects,* which can be extracted using specific object detectors and give valuable hints for the detection of other object types (Hinz and Baumgartner, 2003). If statistical methods are applied, one strategy for incorporating context in the classification process is to augment the feature vectors of each primitive to be classified by additional *context features* (Wolf and Bileschi, 2006) that take into account information from neighbouring primitives. The classification itself can nevertheless be carried out independently for each of the (augmented) feature vectors. However, this may lead to very high-dimensional feature vectors.

As an alternative, the mutual dependencies between neighbouring primitives can be modelled explicitly. In a probabilistic framework, such an approach leads to *graphical models,* in particular to Markov random fields (MRFs) or conditional random fields (CRFs). The remainder of this section will be focused on such approaches. We start with the introduction of some basic principles of graphical models before discussing the MRF and CRF models in detail. This will be followed by an overview about methods for training and for the simultaneous classification of all primitives on the basis of such models, and by some concluding remarks about further developments in context-based classification.

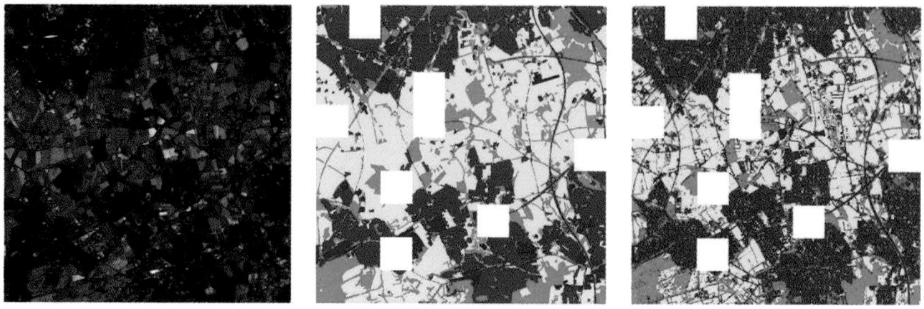

(a) (b) (c)

Figure 5.1 Classification of a satellite image (ground sampling distance: 4 m; area covered by the image: 8.6×8.6 km^2): (a) satellite image; (b) reference; (c) maximum likelihood classification. Classes: *residential* (red), *industrial* (blue), *forest* (green), *cropland* (yellow). White: data used for training. Overal accuracy: 73.2%.

5.1 Graphical Models

Graphical models are a useful tool to visualise the structure of probabilistic problems involving random variables that are mutually dependent on each other. Their application is not limited to classification problems – a survey for applications in geodesy can be found in (Förstner, 2013). The *nodes* of the underlying graph of such a model correspond to the random variables and the fixed parameters of the probabilistic model (cf. Figure 5.2). In this figure random variables are represented by circles with the variable names shown in the circle centres. If a random variable is observed, that is, if a sample of that random variable is available, the corresponding circle is shaded (e.g. y_1 in Figure 5.2). Fixed parameters, such as parameters of prior distributions (*hyperparameters*), are represented by small black circles (e.g. p in Figure 5.2). The *edges* of the graph correspond to dependencies between the variables. Edges can be directed if one can distinguish a *cause* from an *effect* in this relation. Directed edges are represented by arrows pointing from the cause to the effect. In contrast, undirected edges correspond to a generic mutual dependency between two random variables. A graphical model is referred to as a *directed graphical model* if its underlying graph only contains directed edges; otherwise it is called an *undirected graphical model* (Bishop, 2006). A graphical model per se does not contain any information about probability distributions. Nevertheless, there are two major reasons why graphical models have turned out to be very useful for solving probabilistic problems:

- They show the *structure* of the model. In particular, they can be used immediately to derive a *factorisation* of the joint probability distribution of all random variables contained in the model. Using this factorisation, only dependencies that are a part of the graphical model need to be considered. This leads to simplified models for the joint distribution that require a smaller number of parameters than a generic model.

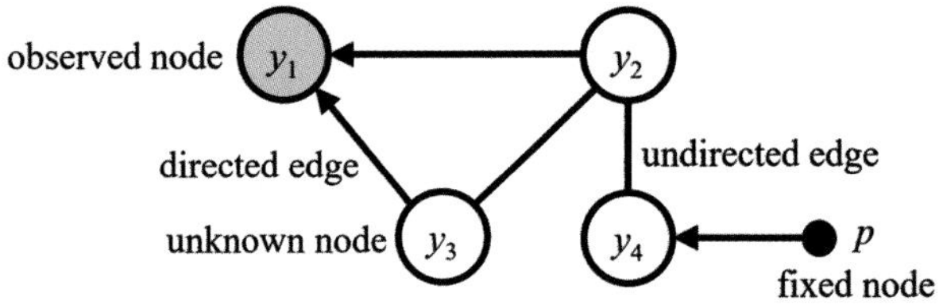

Figure 5.2 A graphical model with different types of nodes and edges.

- For some specific graph structures there are efficient and proven algorithms for the determination of the optimal values of all unknown variables. If an application can be mapped to such a graph, the related optimisation algorithms can be applied as well, independently from the specific type of application.

One important group of graphical models comprises *Bayesian networks.* A Bayesian network is based on a directed graph without cycles (an example is shown in Figure 5.3). In a Bayesian network consisting of N random variables $(y_1, \ldots, y_N)$, the joint distribution of these variables can be factorised according to Equation 5.1 (Bishop, 2006):

$$p\left(y_1, \ldots, y_N\right) = \prod_{i \in \{1,\ldots,N\}} p\left(y_i \mid pa\left(y_i\right)\right). \tag{5.1}$$

In Equation 5.1, $p\left(y_i \mid pa\left(y_i\right)\right)$ denotes the conditional probability of y_i given the states of all of its *parent nodes,* observing the convention $p\left(y_i \mid pa\left(y_i\right)\right) = p\left(y_i\right)$ for nodes without parents. For the example in Figure 5.3 this leads to

$$p\left(y_1, y_2, y_3, y_4\right) = p\left(y_1\right) \cdot p\left(y_2\right) \cdot p\left(y_3 \mid y_1, y_2\right) \cdot p\left(y_4 \mid y_3\right). \tag{5.2}$$

For some types of Bayesian networks, in particular for chains, there are efficient exact techniques for computing the marginal probabilities for all variables in the network (*sum-product* algorithm) and for *inference,* that is, to determine the joint optimal state of all variables (*max-product* algorithm). Both are so-called *message passing* algorithms in which the variables communicate via "messages" that depend on the factors of the joint distribution. In both cases it is sufficient to traverse the graph twice in order to obtain a solution for all variables (Bishop, 2006).

For undirected graphs, factorisation of the joint distribution cannot be directly from a graph in the way discussed up to now. In this case, the factorisation has to be made explicit in the graph by adding an additional node type, the so-called *factor nodes.* Factor nodes are represented by small squares; Figure 5.3 shows an example for a resultant *factor graph.* Each factor node f_F with $F \in \{1, \ldots, N_F\}$ corresponds to a function $f_F\left(\mathbf{y}_F\right)$, where $\mathbf{y}_F \subseteq \{y_1, \ldots, y_N\}$ is the set of variables that serve as the arguments of $f_F\left(\mathbf{y}_F\right)$. The factor node f_F

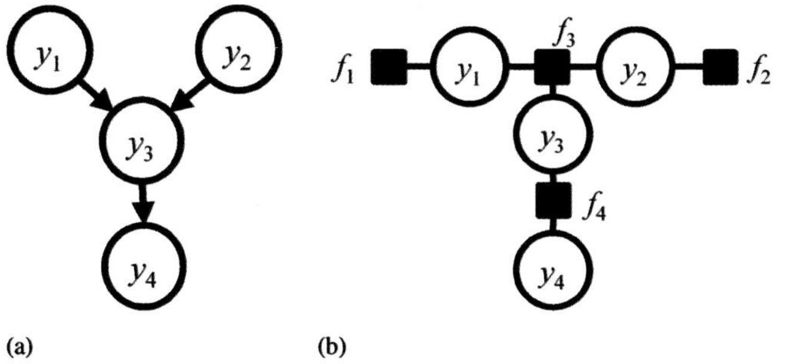

Figure 5.3 Examples for graphical models: (a) Bayesian network; (b) factor graph (adapted from Bishop, 2006).

is connected to all elements of $\mathbf{y}_F$ by an undirected edge. The joint distribution of all random variables in the graph can be modelled by the product of the functions $f_F\left(\mathbf{y}_F\right)$ of all factor nodes f_F (Bishop, 2006):

$$p\left(y_1,\ldots,y_N\right)=\prod_{F\in\{1,\ldots,N_F\}} f_F\left(\mathbf{y}_F\right). \tag{5.3}$$

For the factor graph in Figure 5.3, this leads to

$$p\left(y_1,y_2,y_3,y_4\right)=f_1\left(y_1\right)\cdot f_2\left(y_2\right)\cdot f_3\left(y_1,y_2,y_3\right)\cdot f_4\left(y_3,y_4\right). \tag{5.4}$$

Note that any graph can be represented by a factor graph. The example in Figure 5.3 is chosen so that the factor graph in Figure 5.3b corresponds to the Bayesian network in Figure 5.3a. As a consequence, the factors in Equations 5.2 and 5.4 correspond to each other. However, Equation 5.4 (and the factor graph) no longer indicates which type of (conditional or joint) probability is related to each of the factors.

For factor graphs that can be arranged as trees there are also efficient exact message passing algorithms for the determination of marginal probabilities or for inference (Bishop, 2006). However, for graphs with cycles one has to resort to approximative techniques for the determination of the optimal state of all variables.

The probabilistic classification of images or point clouds can also be described by a graphical model. For the pixel-based classification of an image, such a graphical model is shown in Figure 5.4, which shows the class labels l_i of four neighbouring pixels of an image grid and the corresponding feature vectors $\mathbf{x}_i$. As the features are observed by the sensor, the corresponding nodes are shaded in the figure. The direction of the edges is explained by the fact that the observed features are "caused" by the object properties and not vice versa. This model does not contain dependencies between class labels and/or feature vectors at neighbouring pixels. The graphical model in Figure 5.4 is a Bayesian network, and so the joint probability $p\left(\mathbf{l},\mathbf{x}\right)$ of all class labels and all data can be factorised

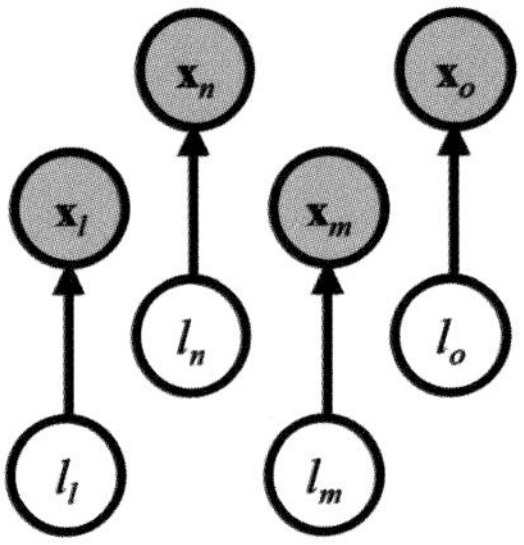

Figure 5.4 Graphical model for the independent classification, using the example of four neighbouring image pixels.

according to Equation 5.1:

$$p(\mathbf{l}, \mathbf{x}) = \prod_{i \in \{1,\ldots,N_P\}} p(\mathbf{x}_i \mid l_i) \cdot p(l_i). \tag{5.5}$$

In Equation 5.5, the vector $\mathbf{l}$ collects the class labels of all image primitives, that is, $\mathbf{l} = (l_1, \cdots, l_N)$, where l_i is the class of primitive i and N is the number of primitives to be classfied, and the notation $\mathbf{x}$ is used to represent *all* data observed in the image. The maximum a posteriori (MAP) principle requires $\mathbf{l}$ to be determined so that the probability in Equation 5.5 is maximised. According to the graph structure, which does not model any dependencies between class labels and data at neighbouring pixels, the product in Equation 5.5 becomes a maximum if each of the factors $p(\mathbf{x}_i \mid l_i) \cdot p(l_i)$ is maximised. Thus, according to the model in Figure 5.4, the most probable configuration of class labels $\mathbf{l}$ of the entire image is obtained if, for each pixel, the optimum class label is determined locally, that is, independently of all other pixels.

5.2 Markov Random Fields

The model for the independent classification of all image primitives described in the previous section is based on an oversimplification. Objects do have a certain, sometimes large, extent in space, and the class labels of all primitives within an object are identical. Different class labels at neighbouring primitives only occur at object boundaries, so that class change can be considered to be a rare and thus an unlikely event. Consequently, the class labels at neighbouring primitives are correlated, a fact that is neglected in the model according to Figure 5.4. In order to expand the model, the dependencies between the class labels have to be considered. This can be done by modelling the prior for the class label of a primitive i by the conditional probability $p(l_i \mid \mathbf{l} \setminus l_i)$ for the class label l_i given the class labels of all other primitives. An undirected graphical model is called a *Markov random field (MRF)* if this conditional probability observes the Markov property (Besag, 1986; Geman and Geman, 1984). In the

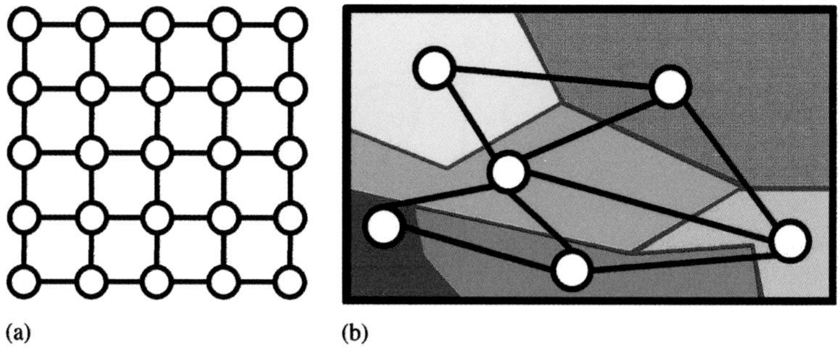

Figure 5.5 Examples for MRF graph structures on images. (a) Nodes: pixels; edges: 4-neighbours on the grid. (b) Nodes: segments; edges: first-order neighbours.

context of classification, the Markov property can be expressed as:

$$p(l_i \mid \mathbf{l} \setminus l_i) = p\left(l_i \mid \mathbf{l}_{N_i}\right). \tag{5.6}$$

where vector $\mathbf{l}_{N_i}$ contains the class labels of all the neighbours of primitive i; that is, class label l_i only depends on the class labels of its neighbours. The set of neighbours is defined according to the underlying graph of the MRF. Figure 5.5 shows two examples for graph structures defined on images. According to the Hammersley–Clifford theorem, the joint probability of all class labels can be described by a Gibbs distribution (Hammersley and Clifford, 1971):

$$p(\mathbf{l}) = \frac{1}{Z} \prod_{k \in \mathbb{K}} \varphi_k(\mathbf{l}_k). \tag{5.7}$$

In Equation 5.7, $\mathbb{K}$ denotes the set of all maximum cliques of the graphical model, $\mathbf{l}_k$ is the set of all class labels contained in clique k, and $\varphi_k(\mathbf{l}_k)$ is the *clique potential* assigned to clique k. Z is a normalising constant called the *partition function*. It is required for the result of Equation 5.7 to be referred to as a probability.

In a Bayesian setting, Equation 5.7 is interpreted as the prior for the classification of all image primitives. In the context of classification, features $\mathbf{x}_i$ observed at individual image primitives i are frequently considered to be conditionally independent. Such a model is called a hidden Markov model (HMM); the underlying graph for the classification of an image is shown in Figure 5.6a. In the example in the figure, the graph edges are based on a four-neighbourhood on the image grid. The term HMM indicates that the unknown variables (in the case of a classification task: the class labels) are not observed (hence "hidden").

The mutual dependencies between neighbouring image primitives and, consequently, the context model are described by the clique potentials in Equation 5.7, which corresponds to the most general case of a MRF. In many applications, the model according to Equation 5.7 is restricted to cliques of order two, that is, to cliques only corresponding to pairs of neighbouring nodes

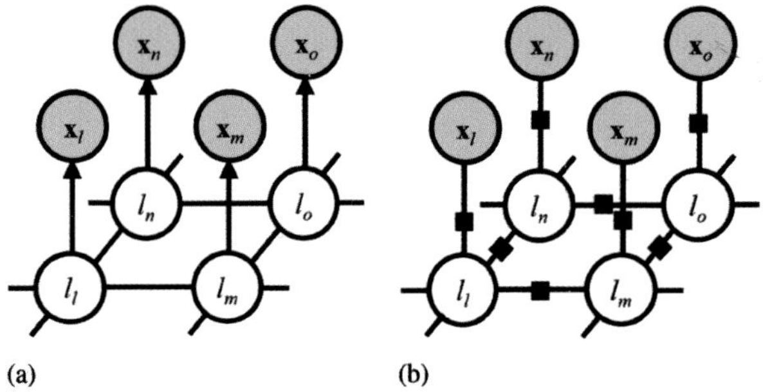

Figure 5.6 (a) Structure of HMM whose nodes correspond to pixels. (b) Corresponding factor graph.

in the graph. This definition is the basis of so-called *pairwise models.* In this case, the joint posterior $p(\mathbf{l} \mid \mathbf{x})$ for all class labels given all the data $\mathbf{x}$ can be factorized according to the factor graph in Figure 5.6a (Besag, 1986; Geman and Geman, 1984; Li, 2009):

$$p(\mathbf{l} \mid \mathbf{x}) = \frac{1}{Z} \cdot \prod_{i \in \mathbb{P}} \varphi_i\left(l_i, \mathbf{x}_i\right) \cdot \prod_{\{i,j\} \in \mathbb{E}} \psi_{ij}\left(l_i, l_j\right). \tag{5.8}$$

In Equation 5.8, $\mathbb{P}$ is the set of all image primitives and i is a specific primitive in $\mathbb{P}$. $\mathbb{E}$ denotes the set of all edges between neighbouring primitives, and the tuple $\{i, j\}$ corresponds to one such edge. Z is the normalisation constant defined previously. In addition, there are two groups of terms in Equation 5.8:

- $\varphi_i\left(l_i, \mathbf{x}_i\right)$. Each of these terms connects the class label of primitive i to the data $\mathbf{x}_i$ observed at that primitive. These terms are called *unary potentials* (because they are only related to one unknown variable) or *node potentials.* One has a lot of freedom in choosing a functional model for these node potentials; in particular, they need not correspond to probabilities. Nevertheless, in a Baysian setting, as has been introduced in this section, it is common to use the likelihood of a local generative classifier to define the node potentials, which leads to

$$\varphi_i\left(l_i, \mathbf{x}_i\right) = p\left(\mathbf{x}_i \mid l_i\right). \tag{5.9}$$

- $\psi_{ij}\left(l_i, l_j\right)$. These terms take as arguments the class labels of two primitives i and j connected by an edge in the graph. They are referred to as *binary potentials* (because they are related to two unknown variables), *pairwise clique potentials* or *edge potentials,* and they define the local context model of the HMM.

Subscripts i in φ_i and ij in ψ_{ij}, respectively, indicate that, in principle, the functional models for the potentials could vary with image position. However, in most applications, the same functional models are used for all primitives. Consequently, these subscripts are neglected in the remainder of this section.

In principle, the pairwise clique potentials can also be chosen freely, and they need not correspond to probabilities. However, they have to be designed in a way to deliver larger values for more likely local configurations of pairs of class labels than for configurations that are less likely. The observation made at the beginning of this section, namely that two neighbouring primitives are more likely to belong to the same class than otherwise, is considered by the two models most frequently used for the pairwise clique potentials (Besag, 1986; Geman and Geman, 1984):

- *Ising model.* This model is applicable for binary classification. Using the class labels $l^1 = +1$ and $l^2 = -1$ to correspond to the foreground and background, respectively, the pairwise clique potential is modelled according to

$$\log \psi \left(l_i, l_j\right) = \beta \cdot l_i \cdot l_j. \tag{5.10}$$

  In Equation 5.10, $\beta > 0$ is a constant modulating the impact of the context model on the classification result. If the class labels are identical, Equation 5.10 results in $\psi \left(l_i, l_j \mid l_i = l_j\right) = e^{\beta} > 1$, whereas for different labels, the pairwise clique potential will be $e^{-\beta} < 1$. Thus, as required, the model according to Equation 5.10 favours label configurations in which neighbouring primitives belong to the same class.
- *Potts model.* This is a generalisation of the Ising model for multi-class problems:

$$\log \psi \left(l_i, l_j\right) = \beta \cdot \delta \left(l_i = l_j\right). \tag{5.11}$$

  In Equation 5.11, $\delta \left(l_i = l_j\right)$ is the Kronecker delta function, giving the result 1 if its argument is true and 0 otherwise. If the class labels are identical, Equation 5.11 results in $\psi \left(l_i, l_j \mid l_i = l_j\right) = e^{\beta} > 1$, whereas for different labels, the pairwise clique potential will be $e^0 = 1$. Again, the model according to Equation 5.11 favours label configurations in which neighbouring primitives belong to the same class.

Compared to an independent classification of all primitives, both models for the pairwise clique potentials will lead to a smoothed classification result with fewer class changes between neighbouring primitives. If the node potential of a primitive does not clearly favour a certain class, the influence of the neighbouring primitives will push the classification result for that primitive towards the class labels of its neighbours.

It is one consequence of the structure of the underlying graph of the HMM that the class labels at neighbouring primitives influence each other mutually. Thus, the optimum configuration of class labels $\mathbf{l}$ according to the MAP criterion can no longer be obtained by an independent classification of each primitive, but the class labels for all primitives are determined simultaneously by maximising

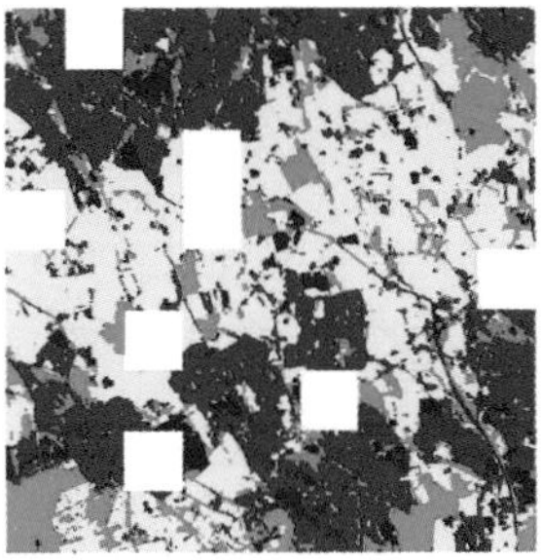

Figure 5.7 Results of the classification of the image shown in Figure 5.1 on the basis of a HMM using the Potts model for pairwise clique potentials. The colours are identical to those in Figure 5.1. The overall accuracy is 78.7%.

the posterior $p(\mathbf{l} \mid \mathbf{x})$ according to the model in Equation 5.8. With few exceptions this is only possible in an approximative way, a statement that also applies to training. An overview about methods for training and inference is given below.

MRFs with cliques consisting of more than two nodes lead to models with higher-order potentials (e.g. Kohli et al., 2009). Using higher-order cliques allows to model more complex dependencies between variables, but training and inference become more difficult as well. This section is restricted to pairwise models.

Figure 5.7 shows the results of an HMM-based classification of the image shown in Figure 5.1. The Potts model was used for the edge potentials in this example. As a consequence of modelling the dependencies between neighbouring class labels, the overall accuracy is increased by 5.5%. Comparing Figures 5.1 and 5.7, it is also obvious that the results of the HMM appear to be smoother. However, in some cases, relevant details are lost in the HMM-based classification, in particular roads (belonging to the class *residential* in the reference). It is a well-known problem of MRF that they tend to oversmooth the result. This problem is caused by the fact that the smoothing effect of the models for the edge potentials according to Equations 5.10 and 5.11 is independent of the local image content.

5.3 Conditional Random Fields

Originally, CRFs were introduced for problems related to text analysis (Lafferty et al., 2001). In (Kumar and Hebert, 2006) they were adapted to problems in image analysis. Just like HMMs, CRFs are based on a model for the posterior $p(\mathbf{l} \mid \mathbf{x})$ of the class label configuration $\mathbf{l}$. Hovever, CRFs are discriminative models, and the posterior $p(\mathbf{l} \mid \mathbf{x})$ is modelled directly (Kumar and Hebert, 2006):

$$p(\mathbf{l} \mid \mathbf{x}) = \frac{1}{Z} \cdot \prod_{i \in \mathbb{P}} \varphi(l_i, \mathbf{x}) \cdot \prod_{\{i,j\} \in \mathbb{E}} \psi(l_i, l_j, \mathbf{x}). \tag{5.12}$$

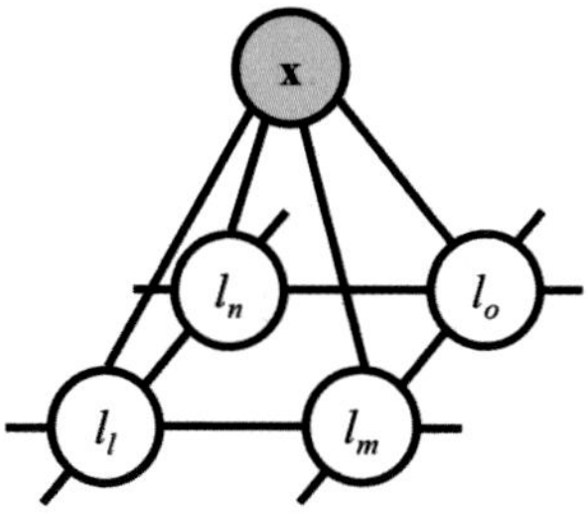

Figure 5.8 Graphical model of a CRF according to Förstner (2013).

The structure of the underlying graph of a CRF is shown in Figure 5.8. The terms in Equation 5.12 correspond to those modelling the posterior of the HMM in Equation 5.8. However, there are two major differences:

1. $\varphi(l_i, \mathbf{x})$. In the context of CRF, the unary potentials are also referred to as *association potentials.* Compared to Equation 5.8, the only difference is that there is no subscript i for the data vector. This implies that in principle the class label of a primitive i may depend on *all available data* $\mathbf{x}$. A common approach to realise this dependency is to represent the data by *node feature vectors* $\mathbf{f}_i(\mathbf{x})$ whose components may be functions of all data $\mathbf{x}$. The second difference from the HMM is that, in general, one uses the probabilistic output of a *discriminative* classifier for these potentials, and consequently, they can be modelled according to

$$\varphi(l_i, \mathbf{x}) = p(l_i \mid \mathbf{f}_i(\mathbf{x})). \tag{5.13}$$

   This means that, for each node i, any discriminative local classifier can be applied to the corresponding node feature vector $\mathbf{f}_i(\mathbf{x})$ to obtain the association potential.
2. $\psi(l_i, l_j, \mathbf{x})$. In the context of CRF, the binary potentials are also referred to as *interaction potentials.* In contrast to the HMM, these potentials also depend on the data. For instance, this dependence can be used to modulate the degree of smoothing according to the local variations of the data.

In order to realise the dependency of the interaction potentials from all data $\mathbf{x}$, the data are represented by *edge feature vectors* $\boldsymbol{\mu}_{ij}(\mathbf{x})$. One such vector is determined for each edge $\{i, j\}$, and its components may (in principle) be functions of all the data. There are several options for defining the edge feature vectors:

- *Difference of the node features,* $\boldsymbol{\mu}_{ij}(\mathbf{x}) = \mathbf{f}_i(\mathbf{x}) - \mathbf{f}_j(\mathbf{x})$. Some of the node features may have been determined from large local neighbourhoods that may overlap considerably for two neighbouring nodes i and j. Consequently, these features will take similar values for two neighbouring nodes, so it may be useful to select a few components of the feature vectors to define the the edge features. The most simple definition is to take the differences of the *features directly observed at the neighbouring nodes,* i.e. $\boldsymbol{\mu}_{ij}(\mathbf{x}) = \mathbf{x}_i - \mathbf{x}_j$.

- *Combination of node features.* In some applications, the edge features are obtained by concatenating the node feature vectors of the two neighbouring primitives, that is, $\boldsymbol{\mu}_{ij}(\mathbf{x}) = [\mathbf{f}_i^T(\mathbf{x}),\ \mathbf{f}_j^T(\mathbf{x})]^T$.
- *Specific edge features.* The edge features need not depend on the node features at all. Thus, arbitrary functions of the data can be used to define edge features to represent the data in the interaction potentials.

An expansion of the Potts model described in the previous section (Equation 5.11) is the *contrast-sensitve Potts model* (Boykov and Jolly, 2001), which is frequently used to model the interaction potentials in CRF:

$$\log \psi\left(l_i, l_j, \mathbf{x}\right) = \delta\left(l_i = l_j\right) \cdot \beta_0 \cdot \left(\beta_1 + (1 - \beta_1) \cdot e^{-\frac{\|\mathbf{x}_i - \mathbf{x}_j\|^2}{2 \cdot \sigma^2}}\right). \tag{5.14}$$

Similar to the Potts model it has a smoothing effect on the classification results. However, the degree of smoothing depends on the difference of the features observed at neighbouring primitives. Parameter $\beta_0 > 0$ corresponds to the weight of the interaction potential relative to the association potential, whereas β_1 with $0 < \beta_1 < 1$ modulates the relative impact of the data-dependent term in Equation 5.14. Parameter σ can be determined as the average squared distance of the features at neighbouring primitives in the training data. If the features observed at two neighbouring primitives are identical, the data-dependent term in Equation 5.14 will tend towards 1, and the interaction potential will have a smoothing influence with weight β_0. If the difference between the two node feature vectors is large, the data-dependent term will tend towards 0. In this case, the interaction potentials will still have a smoothing effect, but the weight of this effect will be $\beta_0 \cdot \beta_1$ and, thus, smaller than β_0 due to $\beta_1 < 1$. Thus, the contrast-sensitive Potts model will lead to a strong degree of smoothing in homogeneous image regions and to a lower degree of smoothing in areas where there are large variations in the data, indicating that a class change is more likely than in homogeneous regions.

As an alternative, the dependency of the interaction potentials from the data can also be modelled by applying an *arbitrary discriminative classifier* with a probabilistic output for that purpose. This can be carried out in two ways (Kumar and Hebert, 2006):

1. *Binary classification.* Based on the interaction features, the classifier decides whether the class labels at two neighbouring primitives are identical or not. The interaction potential is defined on the basis of the posterior

delivered by the classifier:

$$\psi\left(l_i, l_j, \mathbf{x}\right) = \left[p\left(l_i = l_j \mid \boldsymbol{\mu}_{ij}\left(\mathbf{x}\right)\right)\right]^{w}. \tag{5.15}$$

2. *Determination of the optimum local configuration of classes.* Based on the interaction features, the classifier determines the most probable configuration of classes at the two neighbouring primitives. Again, the interaction potential is defined on the basis of the posterior delivered by the classifier:

$$\psi\left(l_i, l_j, \mathbf{x}\right) = \left[p\left(l_i, l_j \mid \boldsymbol{\mu}_{ij}\left(\mathbf{x}\right)\right)\right]^{w}. \tag{5.16}$$

In both cases, $w > 0$ is a weight parameter defining the relative impact of the interaction potential compared to the association potential. Any local discriminative classifier delivering a probabilistic output can be used to define the models according to Equations 5.15 and 5.16. Such models can consider more complex dependencies between the local configuration of class labels and the data than the contrast-sensitive Potts model. However, they require a sufficient amount of *training samples for class transitions* in order to train the underlying classifiers. The generation of such training data is relatively time-consuming and in general requires *fully labelled training image patches.* For the model according to Equation 5.16, in a classification problem involving L different class labels, there are $L \times L$ possible class transitions, and for each of these transitions it is necessary to have a sufficiently large number of training samples. Having to provide these training data may be prohibitive for applying such a model.

Figure 5.9 shows the effects of using a CRF on the basis of a contrast-sensitive Potts model. The figure shows the classification results for the image also used in Figures 5.1 and 5.7. Compared to the classification based on the HMM, the overall accuracy is only increased by a small margin (+0.1%), although it is still considerably better than in the case of an independent pixel-based classification. The resultant label image is less noisy than the one in Figure 5.1, obtained by an independent classification, but it contains more detail than the result of a

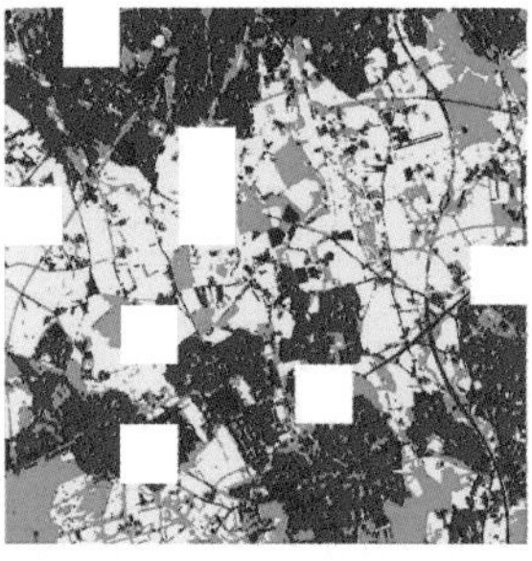

Figure 5.9 Results of the classification of the image shown in Figure 5.1 on the basis of a CRF using the contrast-sensitive Potts model for the interaction potentials. The colours are identical to those in Figure 5.1. The overall accuracy is 78.9%.

classification based on the HMM in Figure 5.7. This preservation of details such as roads, which can still be recognised in Figure 5.9, comes at the cost of a larger rate of errors in areas corresponding to *cropland*, mostly in the form of compact clusters of pixels assigned to a wrong class. However, it is relatively simple to remove such errors in a post-processing state, whereas details removed due to oversmoothing cannot be retrieved easily by post-processing steps.

CRFs do not constitute a specific classification method, but they offer a framework that can be useful for a large variety of applications. In addition, CRFs can be expanded easily, for example by adding different layers of class labels in hierarchical models (Yang and Förstner, 2011). Applications of CRFs differ in the definitions of the graph structure and of the (node and edge) features, as well as in the models used for the association and interaction potentials. Examples from remote sensing include the context-based classification of airborne (Niemeyer et al., 2014) and terrestrial (Weinmann et al., 2015) point clouds, the simultaneous determination of land cover and land use from aerial image and height data (Albert et al., 2017), multi-temporal classification of satellite imagery (Hoberg et al., 2015) and multi-layer CRF for considering occlusions (Kosov et al., 2013).

5.4 Inference and Training

Using MRFs or CRFs, the class labels of all primitives are determined simultaneously by finding the configuration of class labels for which the posterior $p(\mathbf{l} \mid \mathbf{x})$ is maximised. A prerequisite for classification is that the parameters of all the potentials required for modelling $p(\mathbf{l} \mid \mathbf{x})$ are known. These aspects will be discussed briefly in this section.

5.4.1 Inference

An incomplete list of inference techniques could include the following (all except one only provide approximate solutions):

- *Iterative conditional mode (ICM)* (Besag, 1986). This method starts with a local classification using only unary potentials. After that, the optimal class is determined for each primitive, considering the binary potentials and the current optimal state of its neighbours in addition to its unary potentials. This procedure is repeated until a stopping criterion is fulfilled. ICM is considered to be susceptible to local optima.
- *Simulated annealing* (Li, 2009). This is a generative approach based on stochastical sampling.
- *Graph cuts for binary problems* (Boykov et al., 2001). This is the only exact inference technique for MRF/CRF. For that purpose, the graph is expanded by two additional *terminal nodes* called the *source* and the *sink*, which correspond to the two classes to be discerned. In addition to its

neighbours, each image primitive is connected to both terminal nodes by an edge. Instead of potentials, "costs" (corresponding to the inverse potentials) are introduced for all edges of the augmented graph, where the costs of the edges connecting neighbouring primitives are derived from the binary potentials, whereas the costs of the edges connecting an image primitive to the terminal nodes are derived from the unary potential of that primitive. The augmented graph is split into two disjunct parts so that the sum of costs associated to the edges that must be removed for that purpose is minimised. The resultant subgraphs contain the primitives associated to the respective classes. This technique can only be applied if the binary potentials satisfy the *submodularity constraint*:

$$\log \psi(-1,+1) + \log \psi(+1,-1) \leq \log \psi(-1,-1) + \log \psi(+1,+1)\,. \tag{5.17}$$

This can be guaranteed for the Ising, Potts and contrast-sensitive Potts models, but not for an arbitrary classifier as is used, for instance, in Equation 5.15.

- *Graph cuts for multi-class problems* (Boykov et al., 2001). In this case, the multi-class problem has to be mapped to a series of binary problems that can be solved by graph cuts. However, the joint multi-class solution thus obtained is no longer exact. As in the binary case, the binary potentials must be submodular. In the variant *α-expansion*, each image primitive may change its class label to the value α or not; this binary problem has to be solved for each value α (i.e. for each class) in each iteration of the procedure. The technique involving $\alpha - \beta$*-swaps* requires a binary classification to decide for each primitive whose current label is α whether its label should be swapped to β or not. Again, this has to be repeated for each possible pair of labels (α, β) in one iteration. Graph cuts are considered to be one of the best inference methods, but they are not applicable to all problems due to the restrictions imposed by the submodularity constraint.
- *Loopy belief propagation (LBP).* This is an expansion of the message passing techniques for finding the optimal configuration of random variables on tree-structured graphs mentioned previously that can be applied to graphs with cycles. First, the nodes send messages to all their neighbouring edges, and then they collect all the incoming messages from their edges to update their *beliefs*. In the sum-product variant, the *belief* corresponds to the marginal distribution of the class labels of the corresponding primitive (Frey and MacKay, 1998). In the max-product variant (Kolmogorov, 2006), it approximates the posterior distribution. There is no guarantee for LBP to converge. Nevertheless, LBP is considered to be a good inference method. Compared to graph-cuts, its main advantage is that it can also be applied to problems with binary potentials that are not submodular.

5.4.2 Training

It is the goal of training to determine all parameters of the model for the posterior $p(\mathbf{l} \mid \mathbf{x})$ using a set of training data for which both the features and class labels are known. The parameters to be determined comprise the parameters of the unary potentials, $\boldsymbol{\theta}_u$, and those of the binary potentials, $\boldsymbol{\theta}_b$, which are collected in a joint parameter vector $\boldsymbol{\theta} = \left(\boldsymbol{\theta}_u^T, \boldsymbol{\theta}_b^T\right)^T$. Often, the number of parameters in $\boldsymbol{\theta}_b$ is small. For instance, the Ising model (Equation 5.10) only requires the definition of a single parameter β.

First of all, the parameters of both types of potentials can be learned *independently* from each other (Shotton et al., 2009). In this case, the classifier at the basis of the unary potentials is trained in the same way as it would be learned in the case of an independent classification of the image primitives. If the binary potential is also based on a classifier, the latter also has to be trained using training samples, a process that requires a sufficient amount of training samples for all possible class transitions, as mentioned previously. The remaining parameters (weight w of the binary potential if a classifier is used or the parameters of the Ising, Potts or contrast-sensitive Potts models) can be determined by a procedure such as cross validation. For that purpose, a part of the training samples is not used for training of the individual classifiers. After that, the resultant classifier is applied to the remaining training samples, using different values in a given range for the remaining parameters to be determined. The parameter values for which the resulting classification error becomes a minimum is used for the classification of new data.

For the *joint training* of the parameters, the posterior to be maximised in the classification is interpreted as a function $p(\mathbf{l} \mid \mathbf{x}, \boldsymbol{\theta})$ of the parameters $\boldsymbol{\theta}$. The estimation of $\boldsymbol{\theta}$ according to the MAP criterion requires maximisation of the posterior for $\boldsymbol{\theta}$ given the training data, the latter consisting of N_T fully labelled training areas with observations $\mathbf{x}_{T_t}$ and corresponding class labels $\mathbf{l}_{T_t}$:

$$p\left(\boldsymbol{\theta} \mid \mathbf{x}_{T_1}, \ldots, \mathbf{x}_{T_{N_t}}, \mathbf{l}_{T_1} \ldots, \mathbf{l}_{T_{N_t}}\right) \propto p(\boldsymbol{\theta}) \cdot \prod_{t \in \{1,\ldots,N_t\}} p\left(\mathbf{l}_{T_t} \mid \mathbf{x}_{T_t}, \boldsymbol{\theta}\right) \rightarrow max. \tag{5.18}$$

The factorisation over the training areas in Equation 5.18 is due to the assumption that the training areas are statistically independent. The model to be used for classification according to Equations 5.8 or 5.12 can be substituted for $p\left(\mathbf{l}_{T_t} \mid \mathbf{x}_{T_t}, \boldsymbol{\theta}\right)$. The prior $p(\boldsymbol{\theta})$ is required for regularisation; a frequent assumption is a Gaussian distribution with expectation 0 and standard deviation σ for each parameter in $\boldsymbol{\theta}$ (Vishwanathan et al., 2006). Neglecting the prior leads to a maximum likelihood estimation of the parameters. In any case, the estimation of the parameters can only be carried out in an approximative manner, the main reason being that the evaluation of the partition function Z in Equations 5.8 and 5.12 is computationally intractable. One option for optimisation is given by the *pseudo-likelihood approximation* for $p\left(\mathbf{l}_{T_t} \mid \mathbf{x}_{T_t}, \boldsymbol{\theta}\right)$ (Kumar and Hebert, 2006):

$$p\left(\mathbf{l}_{T_t} \mid \mathbf{x}_{T_t}, \boldsymbol{\theta}\right) = \prod_{i \in \mathbb{P}_i} p\left(l_{T_{ti}} \mid \mathbf{x}_{T_t}, \mathbf{l}_{T_{tN_i}}, \boldsymbol{\theta}\right), \tag{5.19}$$

where $l_{T_{ti}}$ is the class label of primitive i in training area t and $\mathbf{l}_{T_{tN_i}}$ denotes the class labels of all neighbours of that primitive. Using this approximation, optimal values for $\boldsymbol{\theta}$ can be determined using standard techniques for optimisation. One example is the limited memory Broyden–Fletcher–Goldfarb–Shanno (LBFGS) method, a pseudo-Newton technique requiring estimation of the gradient and the current value of the target function in each iteration (Liu and Nocedal, 1989). In (Vishwanathan et al., 2006), the LBFGS technique is applied for training CRFs based on a linear model for the association potential.

5.5 Concluding Remarks

The consideration of local context in the classification process can improve the accuracy of the results considerably, in particular if the local neighbourhood used for computing the node features is relatively small. The most obvious effect is an improvement of the accuracy of classes that do not occur very frequently in the data (e.g. Niemeyer et al., 2014). However, the techniques presented here are restricted to modelling interactions at a very small local scale. A more global view on context would be desirable, but it is hard to obtain using pairwise models as introduced here. This is referred to as the problem of *long range interactions*. Another problem of pairwise MRF or CRF is that it is difficult to incorporate model knowledge about the typical shape of objects. There are several approaches for solving these (and related) problems:

- *Hierarchical CRF.* Such models build a graph in different scales, so that long-range interactions can be considered by local interactions at coarse scales. An application of this strategy for the classification of facade images is presented in (Yang and Förstner, 2011).
- *Models for the relative positions of objects.* In (Gould et al., 2008), a classification technique consisting of two steps is presented. The procedure starts with an independent classification of segments. The results of this classification are used to learn probability distributions for the relative positions of objects in an image, which are introduced in a second, CRF-based, classification procedure to serve as a *relative location prior.*
- *Higher-order potentials.* In order to go beyond pairwise models, higher-order cliques can be added to the graph, leading to higher-order potentials connecting all nodes in such a clique. For instance, the *robust P^N Potts model* connects pixels belonging to the same segment with such a potential. The main role of this potential is to penalise configurations in which too many pixels do not belong to the majority class of a segment (Kohli

et al., 2009). In (Wegner et al., 2013) this method is adapted to the problem of detecting roads in aerial imagery.

- *3D models and scene flow.* In (Menze et al., 2015), the determination of *scene flow* between two temporal neighbours in a sequence of stereo image pairs is combined with 3D CAD models of vehicles in order to identify moving objects. The "class label" determined in this process encodes whether an image primitive corresponds to a specific instance of an object of the given type. For that purpose, the shape and pose of each object as well as an additional term for the consistency of each object hypothesis with the scene geometry has to be added to the CRF.
- *Marked point processes.* These can be used to include a strong model about object shape in a pixel-based classification process. Marked point processes provide models for the distribution of configurations of objects of a given type in an image. The optimal configuration of objects given the data is determined by statistical sampling based on simulated annealing. The energy function to be optimised in this context also consists of a data term and an interaction term, frequently based on heuristic assumptions about the appearance of the objects in the data. In (Ortner et al., 2008), marked point processes are used to detect buildings with a rectangular footprint in a digital surface model. In (Chai et al., 2013), the same framework is used to detect a street network. In this application, a specific way for constructing the object ensures that a consistent network topology for the resultant roads is preserved. The combination of marked point processes with a CRF for building detection is described in (Chai et al., 2012).

In any case one has to agree with (Schindler, 2012), who states that for the classification of remote sensing data one should favour methods such as CRF and MRF that go beyound a purely local solution and try to smooth the results at least at a local level.

References

Albert, L., Rottensteiner, F., and Heipke, C. (2017). A higher order Conditional Random Field model for simultaneous classification of land cover and land use, *ISPRS Journal of Photogrammetry and Remote Sensing,* Volume 130, August 2017, pp. 63–80.

Besag, J. (1986). On the statistical analysis of dirty pictures. *Journal of the Royal Statistical Society B,* 48(3):259–302.

Bishop, C. M. (2006). *Pattern Recognition and Machine Learning,* 1st edition. Springer, New York, USA.

Boykov, Y. Y. and Jolly, M.-P. (2001). Interactive graph cuts for optimal boundary and region segmentation of objects in *n*-D images. In *Proceedings of the IEEE International Conference on Computer Vision (ICCV),* Vol. 1, pages 105–112.

Boykov, Y. Y., Veksler, O., and Zabih, R. (2001). Fast approximate energy minimization via graph cuts. *IEEE Transactions on Pattern Analysis and Machine Intelligence*, 23(11):1222–1239.

Chai, D., Förstner, W., and Lafarge, F. (2013). Recovering line-networks in images by junction-point processes. In *Proceedings of the IEEE Conference on Computer Vision and Pattern Recognition (CVPR)*, pages 1894–1901.

Chai, D., Förstner, W., and Yang, M. Y. (2012). Combine Markov Random Fields and Marked Point Processes to extract building from remotely sensed image. In *ISPRS Annals of the Photogrammetry, Remote Sensing and Spatial Information Sciences*, Vol. I-3, pages 365–370.

Förstner, W. (2013). Graphical models in geodesy and photogrammetry. *Photogrammetrie Fernerkundung Geoinformation*, 4/2013:255–268.

Frey, B. J. and MacKay, D. J. C. (1998). A revolution: belief propagation in graphs with cycles. In *Proceedings of the Neural Information Processing Systems Conference*, pages 479–485.

Geman, S. and Geman, D. (1984). Stochastic relaxation, Gibbs distributions, and the Bayesian restoration of images. *IEEE Transactions on Pattern Analysis and Machine Intelligence*, 6(6):721–741.

Gould, S., Rodgers, J., Cohen, D., Elidan, G., and Koller, D. (2008). Multi-class segmentation with relative location prior. *International Journal of Computer Vision*, 80(3):300–316.

Hammersley, J. M. and Clifford, P. (1971). Markov fields on finite graphs and lattices. Unpublished manuscript, available at http://www.statslab.cam.ac.uk/ grg/books/hammfest/hamm-cliff.pdf (accessed 14 September 2016).

Hinz, S. and Baumgartner, A. (2003). Automatic extraction of urban road networks from multi-view aerial imagery. *ISPRS Journal of Photogrammetry and Remote Sensing*, 58(1–2):83–98.

Hoberg, T., Rottensteiner, F., Feitosa, R. Q., and Heipke, C. (2015). Conditional Random Fields for multitemporal and multiscale classification of optical satellite imagery. *IEEE Transactions on Geoscience and Remote Sensing*, 53(2):659–673.

Kohli, P., Ladický, L., and Torr, P. (2009). Robust higher order potentials for enforcing label consistency. *International Journal of Computer Vision*, 82(3):302–324.

Kolmogorov, V. (2006). Convergent tree-reweighted message passing for energy minimization. *IEEE Transactions on Pattern Analysis and Machine Intelligence*, 28(10):1568–1583.

Kosov, S., Rottensteiner, F., and Heipke, C. (2013). Sequential Gaussian mixture models for two-level conditional random fields. In *Proceedings of the 35th German Conference on Pattern Recognition (GCPR)*, Vol. 8142 of *LNCS*, pages 153–163, Heidelberg. Springer.

Kumar, S. and Hebert, M. (2006). Discriminative Random Fields. *International Journal of Computer Vision*, 68(2):179–201.

Lafferty, J., McCallum, A., and Pereira, F. (2001). Conditional random fields: Probabilistic models for segmenting and labeling sequence data. In *Proceedings of the 18th International Conference on Machine Learning*, pages 282–289.

Li, S. Z. (2009). *Markov Random Field modeling in image analysis*, 3rd edition. Springer, London, UK.

Liu, D. and Nocedal, J. (1989). On the limited memory BFGS method for large scale optimization. *Mathematical Programming*, 45(1–3):503–528.

Menze, M., Heipke, C., and Geiger, A. (2015). Joint 3D estimation of vehicles and scene flow. In *ISPRS Annals of the Photogrammetry, Remote Sensing and Spatial Information Sciences*, Vol. II-3/W5, pages 427–434.

Niemeyer, J., Rottensteiner, F., and Sörgel, U. (2014). Contextual classification of lidar data and building object detection in urban areas. *ISPRS Journal of Photogrammetry and Remote Sensing*, 87(2014):152–165.

Ortner, M., Descombe, X., and Zerubia, J. (2008). A marked point process of rectangles and segments for automatic analysis of digital elevation models. *IEEE Transactions on Pattern Analysis and Machine Intelligence*, 30(1):105–119.

Schindler, K. (2012). An overview and comparison of smooth labeling methods for land-cover classification. *IEEE Transactions on Geoscience and Remote Sensing*, 50(11):4534–4545.

Shotton, J., Winn, J., Rother, C., and Criminisi, A. (2009). Textonboost for image understanding: Multi-class object recognition and segmentation by jointly modeling texture, layout, and context. *International Journal of Computer Vision*, 81(1):2–23.

Vishwanathan, S. V. N., Schraudolph, N. N., Schmidt, M. W., and Murphy, K. P. (2006). Accelerated training of Conditional Random Fields with stochastic gradient methods. In *Proceedings of the 23rd International Conference on Machine Learning*, pages 969–976.

Wegner, J. D., Montoya-Zegarra, J. A., and Schindler, K. (2013). A higher-order CRF model for road network extraction. In *Proceedings of the IEEE Conference on Computer Vision and Pattern Recognition (CVPR)*, pages 1698–1705.

Weinmann, M., Schmidt, A., Mallet, C., Hinz, S., Rottensteiner, F., and Jutzi, B. (2015). Contextual classification of point cloud data by exploiting individual 3D neighbourhoods. In *ISPRS Annals of the Photogrammetry, Remote Sensing and Spatial Information Sciences*, Vol. II-3/W4, pages 271–278.

Wolf, L. and Bileschi, S. (2006). A critical view on context. *International Journal of Computer Vision*, 69(2):251–261.

Yang, M. Y. and Förstner, W. (2011). A hierarchical Conditional Random Field model for labelling and classifying images of man-made scenes. In *Proceedings of the 1st IEEE/ISPRS Workshop on Computer Vision for Remote Sensing of the Environment*, pages 196–203.

Table 6.2 Confusion matrix (binary classification)

	Reference		
Segmentation/	TP	FP	TP + FP
Classification	$\lvert\mathcal{R}_P \cap \mathcal{S}_P\rvert$	$\lvert\mathcal{S}_P \setminus \mathcal{R}_P\rvert$	$\lvert\mathcal{S}_P\rvert$
	FN	TN	TN + FN
	$\lvert\mathcal{R}_P \setminus \mathcal{S}_P\rvert$	$\lvert\mathcal{R}_N \cap \mathcal{S}_N\rvert$	$\lvert\mathcal{S}_N\rvert$
	TP + FN	FP + TN	
	$\lvert\mathcal{R}_P\rvert$	$\lvert\mathcal{R}_N\rvert$	$\lvert\mathcal{I}\rvert$

and for each class the *user accuracy* or *consumer accuracy*

$$UA = \frac{n_{ii}}{n_{i+}} \tag{6.19}$$

and the *producer accuracy*

$$PA = \frac{n_{ii}}{n_{+i}} \tag{6.20}$$

If segments directly form the basis for evaluation, the number n_{ij} is the number of reference segments for which $\mathcal{S}_{\mathcal{R}_{i[k]}} \neq \emptyset$; otherwise n_{ij} denotes the overlapping area following Zhan et al. (2005), but there may be reference segments $\mathcal{R}_{i[k]}$ that are not matched.

In the following we will compile standard quantities for quantitative evaluation. The given publications are publications using these quantitities. For ease of reading, we will first omit the distinction between single segments and consider only a two-class classification for a pixel-level evalution where $\mathcal{R}_P$ denotes reference pixels and $\mathcal{S}_P$ extracted pixels, and the background in an image

$$\mathcal{R}_N = \mathcal{R}_P^* = \mathcal{I} \setminus \mathcal{R}_P \tag{6.21}$$

and

$$\mathcal{S}_N = \mathcal{S}_P^* = \mathcal{I} \setminus \mathcal{S}_P \tag{6.22}$$

respectively. For such a binary classification the confusion matrix reduces to a 2×2 matrix (cf. Table 6.2) and the quantities *true positives* (TP), *true negatives* (TN), *false positives* (FP) and *false negatives* (FN) follow

$$\begin{aligned} TP &= \lvert\mathcal{R}_P \cap \mathcal{S}_P\rvert \quad & TN &= \lvert\mathcal{R}_N \cap \mathcal{S}_N\rvert \\ FP &= \lvert\mathcal{S}_P \setminus \mathcal{R}_P\rvert \quad & FN &= \lvert\mathcal{R}_P \setminus \mathcal{S}_P\rvert \end{aligned} \tag{6.23}$$

All positives P are given by

$$P = TP + FN = \lvert\mathcal{R}_P\rvert \tag{6.24}$$

and all negatives N by

$$N = TN + FP = |\mathcal{R}_N| \tag{6.25}$$

Based on these quantities, a number of quantities with different properties are defined. If we want to just compare two segmentations/classifications $\mathcal{R} = \mathcal{S}_1$ and $\mathcal{S} = \mathcal{S}_2$ – thus an unsupervised evaluation (Chabrier et al., 2006; Corcoran et al., 2010) – an applicable quantity should be defined also in cases where $\mathcal{R} = \emptyset$, $\mathcal{S} = \emptyset$ and $\mathcal{R} \cap \mathcal{S} = \emptyset$. If $\mathcal{R}$ is a reference set – thus a supervised evaluation – then the assumption $\mathcal{R} \neq \emptyset$ holds, but the intersection of the two sets may still be an empty set. For all quantities the range is given and the value for the best result is underlined.

The first four commonly used quantities for evaluation are described in the following by relating the entries of the confusion matrix to the sums of its columns.

- *True positive rate* (*TPR*)

$$TPR = \frac{TP}{P} = \frac{|\mathcal{R}_P \cap \mathcal{S}_P|}{|\mathcal{R}_P|} \tag{6.26}$$

  with $TPR \in [0, \underline{1}]$ and defined for $\mathcal{R}_P \neq \emptyset$. This quantity is also named the *detection rate* ρ_d (McGlone and Shufelt, 1994; McKeown et al., 2000), *recall* (Estrada and Jepson, 2005; Karaman et al., 2005), *sensitivity* (Baldi et al., 2000; Baumann et al., 2008), *completeness* (Heipke et al., 1997; Niederöst, 2001) and relates to the *producer accuracy* given in Equation 6.20.
- *False negative rate* (*FNR*)

$$FNR = \frac{FN}{P} = \frac{|\mathcal{R}_P \setminus \mathcal{S}_P|}{|\mathcal{R}_P|} \tag{6.27}$$

  with $FNR \in [\underline{0}, 1]$ and defined for $\mathcal{R}_P \neq \emptyset$. This quantity is also named the *omission error* ε_o or *classification error type II* β. Furthermore $TPR + FNR = 1$ holds.
- *False positive rate* (*FPR*)

$$FPR = \frac{FP}{N} = \frac{|\mathcal{S}_P \setminus \mathcal{R}_P|}{|\mathcal{R}_N|} = \frac{|\mathcal{S}_P \setminus \mathcal{R}_P|}{|\mathcal{I} \setminus \mathcal{R}_P|} \tag{6.28}$$

  with $FPR \in [\underline{0}, 1]$ and defined if $\mathcal{R} \subset \mathcal{I}$ and thus $\mathcal{R}_N \neq \emptyset$. It is also known as *classification error type I* α. In the case described above where the reference data set is given by the image $\mathcal{I}$ and all positives P and negatives N are well defined, the only problem in interpretation of the quantity arises when the extension of the data set is enlarged without change of the positives P, leading to a lower value of *FPR*. In the general case, it may be difficult to define the set $\mathcal{D} \setminus \mathcal{R}_P$ and thereby the negatives.

- *True negative rate* (*TNR*)

$$TNR = \frac{TN}{N} = \frac{|\mathcal{R}_N \cap \mathcal{S}_N|}{|\mathcal{R}_N|} = \frac{|(\mathcal{I} \setminus R_P) \cap (\mathcal{I} \setminus S_P)|}{|\mathcal{I} \setminus \mathcal{R}_P|} \tag{6.29}$$

  with $TNR \in [0, \underline{1}]$ and $TNR + FPR = 1$. For this quantity the remarks for the FPR again hold and it is also known as *specificity* (Baumann et al., 2008).

The quantities above either consider overlapping or non-overlapping areas, but relate these to $\mathcal{R}_P$ or $\mathcal{R}_N$ and are thereby non-symmetric with regard to $\mathcal{R}$ and $\mathcal{S}$. TPR and TNR are used to define the

- Youden statistic (Youden, 1950):

$$\begin{aligned} Y = TPR + TNR - 1 &= \frac{TP}{P} + \frac{TN}{N} - 1 \\ &= \frac{|\mathcal{R}_N| \cdot |\mathcal{R}_P \cap \mathcal{S}_P| + |\mathcal{R}_P| \cdot |\mathcal{R}_N \cap \mathcal{S}_N| - |\mathcal{R}|}{|\mathcal{R}|} \end{aligned} \tag{6.30}$$

  with $Y \in [0, \underline{1}]$.

Relating the sums of the main diagonal and off-diagonal entries to the number of all elements leads to the following quantities:

- *(Overall) accuracy* ρ_{acc} (c.f. Equation 6.17)

$$\rho_{acc} = \frac{TP + TN}{P + N} = \frac{|\mathcal{R}_P \cap \mathcal{S}_P| + |\mathcal{R}_N \cap \mathcal{S}_N|}{|\mathcal{I}|} \tag{6.31}$$

  with $\rho_{acc} \in [0, \underline{1}]$ and is always defined.
- *Misclassification error* ε

$$\varepsilon = \frac{FP + FN}{P + N} = \frac{|\mathcal{S}_P \setminus \mathcal{R}_P| + |\mathcal{R}_P \setminus \mathcal{S}_P|}{|\mathcal{I}|} \tag{6.32}$$

  with $\varepsilon \in [\underline{0}, 1]$ and $\varepsilon = 1 - \rho_{acc}$. Furthermore the *misclassification error* relates to the *classification error type I* α and the *classification error type II* β by $\varepsilon = (1 - r)\alpha + r\beta$ with $r = \frac{|\mathcal{S}|}{|\mathcal{I}|}$.

The overall accuracy ρ_{acc} and *misclassification error* ε are symmetric with respect to $\mathcal{R}$ and $\mathcal{S}$, because they either consider all overlapping or all non-overlapping elements. The next group of quantities again use single entries of the confusion matrix, but relate these to the sums of its rows.

- *Positive predictive value* (*PPV*)

$$PPV = \frac{TP}{TP + FP} = \frac{|\mathcal{R}_P \cap \mathcal{S}_P|}{|\mathcal{R}_P \cap \mathcal{S}_P| + |\mathcal{S}_P \setminus \mathcal{R}_P|} = \frac{|\mathcal{R}_P \cap \mathcal{S}_P|}{|\mathcal{S}_P|} \tag{6.33}$$

with $PPV \in [0, \underline{1}]$ and defined for $\mathcal{S}_P \neq \emptyset$, which is unlikely for the evalution of a classification of an entire data set, but may occur for single segments dependent on the assignment scheme. This quantity is also known as *precision* (Estrada and Jepson, 2005; Karaman et al., 2005) and *correctness* (Heipke et al., 1997). It relates to the *user accuracy* as given in Equation 6.19 and to the *commission error* (Equation 6.35) by $\varepsilon_c = 1 - PPV$.

- *Negative predictive value* (*NPV*)

$$NPV = \frac{TN}{TN + FN} = \frac{|\mathcal{R}_N \cap \mathcal{S}_N|}{|\mathcal{R}_N \cap \mathcal{S}_N| + |\mathcal{R} \setminus \mathcal{S}|} = \frac{|\mathcal{R}_N \cap \mathcal{S}_N|}{|\mathcal{S}_N|} \tag{6.34}$$

which is defined for $\mathcal{S}_N \neq \emptyset$ and $NPV \in [0, \underline{1}]$.

- *Commission error* ε_c

$$\varepsilon_c = \frac{FP}{TP + FP} = \frac{|\mathcal{S}_P \setminus \mathcal{R}_P|}{|\mathcal{S}_P|} \tag{6.35}$$

with $\varepsilon_c \in [\underline{0}, 1]$ and defined for $\mathcal{S}_P \neq \emptyset$.

The next group of quantities relate the sum of off-diagonal elements to either single elements or sums of selected elements of the confusion matrix aiming at symmetric quantities:

- *Quality rate* ρ_q

$$\rho_q = \frac{TP}{TP + FP + FN} = \frac{|\mathcal{S}_P \cap \mathcal{R}_P|}{|\mathcal{S}_P \cup \mathcal{R}_P|} \tag{6.36}$$

$$= 1 - \frac{FP + FN}{TP + FP + FN} = 1 - \frac{|(\mathcal{S}_P \setminus \mathcal{R}_P) \cup (\mathcal{R}_P \setminus \mathcal{S}_P)|}{|\mathcal{S}_P \cup \mathcal{R}_P|} \tag{6.37}$$

with $\rho_q \in [0, \underline{1}]$. This quantity is also known as the *Jaccard index* and is used in McGlone and Shufelt (1994), Heipke et al. (1997), McKeown et al. (2000) and Ragia et al. (2000).

- *Shape difference* δ_s (Keim, 1999)

$$\delta_s = 1 - \frac{TP}{TP + FP + FN} = 1 - \frac{|\mathcal{S}_P \cap \mathcal{R}_P|}{|\mathcal{S}_P \cup \mathcal{R}_P|} \tag{6.38}$$

with $\delta_s \in [\underline{0}, 1]$ defined for either $\mathcal{R}_P \neq \emptyset$ or $\mathcal{S}_P \neq \emptyset$ and relates to the *quality rate* (Equation 6.36) by $\delta_s = 1 - \rho_q$.

- *Hammoude metric* (Marcal and Rodrigues, 2008)

$$\rho_H = \frac{FP + FN}{TP + FP + FN} = \frac{|\mathcal{S}_P \setminus \mathcal{R}_P| - |\mathcal{R}_P \setminus \mathcal{S}_P|}{|\mathcal{R}_P \cup \mathcal{S}_P|} \tag{6.39}$$

$$= \frac{|\mathcal{R}_P \cup \mathcal{S}_P| - |\mathcal{R}_P \cap \mathcal{S}_P|}{|\mathcal{R}_P \cup \mathcal{S}_P|}$$

with $\rho_H \in [\underline{0}, 1]$ and equals the *shape difference* δ_s.

- *Shape dissimilarity* ρ_s (Henricsson and Baltsavias, 1997; Niederöst, 2001)

$$\rho_s = \frac{FP + FN}{P} = \frac{|\mathcal{S}_P \setminus \mathcal{R}_P| + |\mathcal{R}_P \setminus \mathcal{S}_P|}{|\mathcal{R}_P|} \tag{6.40}$$

with $\rho_s \in [\underline{0}, \infty[$ and defined for $\mathcal{R}_P \neq \emptyset$.

Note that all of these quantities do not take the *true negatives* into account, and that ρ_q, δ_s and ρ_H are symmetric with respect to $\mathcal{R}$ and $\mathcal{S}$.

The next group of quantities use single off-diagonal entries and relate them to either single other entries or the sums of rows and columns leading to non-symmetric non-normed quantities.

- *False alarm rate* (*FAR*) (Ragia et al., 2000)

$$FAR = \frac{FP}{P} = \frac{|\mathcal{S}_P \setminus \mathcal{R}_P|}{|\mathcal{R}_P|} \tag{6.41}$$

which is defined for $\mathcal{R}_P \neq \emptyset$ with $FAR \in [\underline{0}, \infty[$.

- *Branch factor* ρ_b (McKeown et al., 2000; McGlone and Shufelt, 1994)

$$\rho_b = \frac{FP}{TP} = \frac{|\mathcal{S}_P \setminus R_P|}{|\mathcal{R}_P \cap \mathcal{S}_P|} \tag{6.42}$$

defined for $\mathcal{R}_P \cap \mathcal{S}_P \neq \emptyset$ with $\rho_b \in [\underline{0}, \infty[$.

- *Miss factor* ρ_m (McGlone and Shufelt, 1994)

$$\rho_m = \frac{FN}{TP} = \frac{|\mathcal{R}_P \setminus S_P|}{|\mathcal{R}_P \cap \mathcal{S}_P|} \tag{6.43}$$

for which the same remarks as for the *branch factor* hold.

Other symmetric quantitities are based on the overlap of sets and relate the size of overlap to the size of the sets in order to measure similarity.

- *Dice similarity coefficient* (Crum et al., 2006; Pluempitiwiriyawei et al., 2004)

$$\rho_{DSC} = 2 \cdot \frac{|\mathcal{R}_P \cap \mathcal{S}_P|}{|\mathcal{R}_P| + |\mathcal{S}_P|} = \frac{2 \cdot TP}{2 \cdot TP + FP + FN} \tag{6.44}$$

with $\rho_{DSC} \in [\underline{0}, 1]$ again defined for either $\mathcal{R}_P \neq \emptyset$ or $\mathcal{S}_P \neq \emptyset$. The *Dice similarity coefficient* is also given by $\rho_{DSC} = \frac{2\rho_q}{1+\rho_q}$ based on the *quality rate* given in Equation 6.36. It equals the *Sørensen similarity* (Sørensen, 1948) used in ecology (Dalirsefat et al., 2009; Looman and Campbell, 1960).

- *Similarity coefficent*

$$\rho_{smax} = \frac{|\mathcal{R}_P \cap \mathcal{S}_P|}{max(|\mathcal{R}_P|, |\mathcal{S}_P|)} = \frac{TP}{max(TP + FN, TP + FP)} \tag{6.45}$$

with $\rho_{smax} \in [0, \underline{1}]$ and defined for either $\mathcal{R}_P \neq \emptyset$ or $\mathcal{S}_P \neq \emptyset$. The quantity $1 - \rho_{smax}$ is used in Mao et al. (2010).

The last group of quantities given here are based on combining non-symmetric quantities in order to obtain a symmetric quantity for evaluation, e.g. *recall* (Equation 6.26) and *precision* (Equation 6.33).

- *F-measure* F_ω (Baumann et al., 2008)

$$F_\omega = (1 + \omega^2) \frac{precision \cdot recall}{\omega^2 \cdot precision + recall} \tag{6.46}$$

with $F_\omega \in [0, \underline{1}]$.
- *F-score* F_1 follows from Equation 6.46 with $\omega = 1$ (Baumann et al., 2008; Karaman et al., 2005):

$$\begin{aligned} F_1 &= \frac{2 \cdot precision \cdot recall}{precision + recall} = \frac{2 \cdot \frac{TP}{TP+FP} \cdot \frac{TP}{TP+FN}}{\frac{TP}{TP+FP} + \frac{TP}{TP+FN}} \\ &= \frac{2 \cdot \frac{|\mathcal{R}_P \cap \mathcal{S}_P|}{|\mathcal{S}_P|} \cdot \frac{|\mathcal{R}_P \cap \mathcal{S}_P|}{|\mathcal{R}_P|}}{\frac{|\mathcal{R}_P \cap \mathcal{S}_P|}{|\mathcal{S}_P|} + \frac{|\mathcal{R}_P \cap \mathcal{S}_P|}{|\mathcal{R}_P|}} \\ &= 2 \cdot \frac{|\mathcal{R}_P \cap \mathcal{S}_P|^2}{|\mathcal{R}_P| \cdot |\mathcal{S}_P|} : \frac{|\mathcal{R}_P \cap \mathcal{S}_P|(|\mathcal{R}_P| + |\mathcal{S}_P|)}{|\mathcal{R}_P| \cdot |\mathcal{S}_P|} \\ &= 2 \cdot \frac{|\mathcal{R}_P \cap \mathcal{S}_P|}{|\mathcal{R}_P| + |\mathcal{S}_P|} \end{aligned} \tag{6.47}$$

with $F_1 \in [0, \underline{1}]$. This equals the *Dice similarity coefficient* given in Equation 6.44.

For investigations on a segmentation benchmark data set Martin et al. (2001) proposed the following:

- Local consistency error (*LCE*)

$$LCE = \frac{1}{|\mathcal{I}|} \sum_{(x,y) \in \mathcal{I}} max(\varepsilon_\gamma(\mathcal{R}, \mathcal{S}; x, y), \varepsilon_\gamma(\mathcal{S}, \mathcal{R}; x, y)) \tag{6.48}$$

with $LCE \in [\underline{0}, 1]$ and the

- Global consistency error (GCE)

$$GCE \frac{1}{|\mathcal{I}|} min \left(\sum_{(x,y)\in\mathcal{I}} \varepsilon_\gamma(\mathcal{R},\mathcal{S};x,y), \sum_{(x,y)\in\mathcal{I}} \varepsilon_\gamma(\mathcal{S},\mathcal{R};x,y) \right) \tag{6.49}$$

with $GCE \in [\underline{0}, 1]$ based on their local refinement error

$$\varepsilon_\gamma(\mathcal{A},\mathcal{B};x,y) = \frac{|\mathcal{A}(x,y) \setminus \mathcal{B}(x,y)|}{|\mathcal{A}(x,y)|} \tag{6.50}$$

with $\varepsilon_\gamma \in [\underline{0}, 1]$. In the case $\mathcal{A} = \mathcal{R}$ and $\mathcal{B} = \mathcal{S}$, ε_γ relates to the *false negative rate*, otherwise to the *classification error type II* β. LCE and GCE are used for comparisons of segmentations (Jiang et al., 2006; Unnikrishnan et al., 2007). Polak et al. (2009) discuss the properties of these quality measures. Based on the discussion they propose the

- Object-level consistency error (OCE)

$$OCE = min(\varepsilon_\tau(\mathcal{R},\mathcal{S}), \varepsilon_\tau(\mathcal{S},\mathcal{R})) \tag{6.51}$$

with

$$\varepsilon_\tau(\mathcal{A},\mathcal{B}) = \sum_j \left(1 - \sum_i \frac{|\mathcal{A}_j \cap \mathcal{B}_i|}{|\mathcal{A}_j \cup \mathcal{B}_i|} \cdot w_{ij}\right) \cdot w_j \tag{6.52}$$

$$w_{ij} = \frac{\delta(|\mathcal{A}_j \cap \mathcal{B}_i|) \cdot |\mathcal{B}_i|}{\sum_l \delta(|\mathcal{A}_j \cap \mathcal{B}_l|) \cdot |\mathcal{B}_l|} \tag{6.53}$$

$$w_j = \frac{|\mathcal{A}_j|}{\sum_l |\mathcal{A}_l|} \tag{6.54}$$

$$\delta(x) = \begin{cases} 0 & \text{if x = 0} \\ 1 & \text{else} \end{cases} \tag{6.55}$$

with $OCE \in [\underline{0}, 1]$. The inner sum is the summation of the quality rates (Equation 6.36) of overlapping segments of both sets weighted by the amount of their overlap. OCE is symmetric with respect to the two sets and inludes the evaluation of the granularity of the two segmentations by the weights, thus *under-* and *oversegmentation*, which will be discussed in Section 6.3.3.

The requirements for quality evaluation may differ significantly. For most cases, the overlap of the sets $\mathcal{R} \cap \mathcal{S}$ is decisive. Therefore, symmetric criteria taking the *true positives* into account are often combined with non-symmetric criteria considering the *false negatives* and *false positives*. In the case of object detection with the requirement that all objects are detected, the consideration of *false negatives* and *false positives* may have different importance, leading, for example, to the use of the *F-measure* (Equation 6.46), which is symmetric only if $\omega = 1$.

6.3.3 Characterising Under- and Oversegmentation

In order to define quantities for the characterisation of a segmentation in terms of *under-* and *oversegmentation*, we first have to define them. Cho and Meer (1997) describe *undersegmentation* as the partition of an image into large erroneous regions and *oversegmentation* as a partition of an image into too many segments. To decide which criteria are significant for their evaluation we need a more precise notion and formulation.

The terms *under-* and *oversegmentation* are often used, but a closer look reveals that the meaning differs in different publications.

Definition 6.3.1 (using homogeneity): *Given a segmentation $\mathcal{S}$ it is undersegmented if it does not fulfil condition* (6.5), *and it is oversegmented if it does not fulfil condition* (6.6).

This definition is used, for example, by Felzenszwalb and Huttenlocher (1998) and Andres et al. (2009), and solely depends on the segmentation result itself, namely the homogeneity of each segment (6.5) and the non-homogeneity of the union (6.6), implying that *under-* and *oversegmentation* are not only dependent on a single segment's properties, but also on its neighbourhood and therefore on the segmentation $\mathcal{S}$. Consideration of the homogeneity is motivated by the fact that segmentation and classification may rely on different criteria. Therefore, the spectral properties should be evaluated for the segments. The critical case is *undersegmentation* with respect to the (spectral) properties used for classification.

Other definitions are based on a reference to which a segmentation has to be evaluated in order to determine if we do have an *under-* or *oversegmented* result.

Definition 6.3.2 (using overlap): *Given a reference segmentation $\mathcal{R}$ and a segmentation $\mathcal{S}$ the segmentation $\mathcal{S}$ is undersegmented if segments $\mathcal{S}_i$ overlaps parts which do not belong to the related reference segment $\mathcal{R}_j$, and the segmentation $\mathcal{S}$ is oversegmented if segments $\mathcal{S}_i$ do not overlap the entire related reference segment $\mathcal{R}_j$.*

This definition is used, for example, by Clinton et al. (2008) and Persello and Bruzzone (2010). It is closely related to the following definition.

Definition 6.3.3 (using number of segments): *Given a reference segmentation $\mathcal{R}$ and a segmentation $\mathcal{S}$ the segmentation $\mathcal{S}$ is undersegmented if segments of $\mathcal{S}$ are related to more than 1 segment of $\mathcal{R}$. The segmentation $\mathcal{S}$ is oversegmented if segments of $\mathcal{R}$ are related to more than 1 segment of $\mathcal{S}$.*

This is used, for example, by Sonka and Fitzpatrick (2000). Marpu et al. (2010) use the term *objects* in their definition, but an object like a building may consist of a number of segments that are spectrally different and therefore the entire object has to be split into a number of segments to fulfil condition (6.5).

The posibility for quantifying *under-* and *oversegmentation* depends on the used definition. If we base our considerations on Definition 6.3.2, both effects can be quantified locally for each segment by the following quantities (Persello and Bruzzone, 2010):

- Undersegmentation index ($USeg$)

$$USeg = 1 - \frac{|\mathcal{R}_i \cap \mathcal{S}_{R_i}|}{|\mathcal{S}_{R_i}|} \qquad (6.56)$$

  with $USeg \in [\underline{0}, 1]$. It is related to the PPV (Equation 6.33) and is only defined if there is at least one segment $\mathcal{S}_i$ assigned to $\mathcal{R}_j$.
- Oversegmentation index ($OSeg$)

$$OSeg = 1 - \frac{|\mathcal{R}_i \cap \mathcal{S}_{R_i}|}{|\mathcal{R}_i|} \qquad (6.57)$$

  with $OSeg \in [\underline{0}, 1]$. It is related to the TPR (Equation 6.26).

Let us consider the case that, according to the assignment criterion 6.12 used by Persello and Bruzzone (2010), a segment $\mathcal{S}_i$ overlaps a segment $\mathcal{R}_j$ just by one. In this configuration $USeg$ and $OSeg$ would be close to one although it may not be meaningful for evaluation.

Besides these two quantities they propose a fragmentation index:

- Fragmentation index ($FSeg$)

$$FSeg(\mathcal{R}_i, \mathcal{S}_{\mathcal{R}_i}) = \frac{\sharp\mathcal{S}_{\mathcal{R}_i} - 1}{\sharp_P(\mathcal{R}_i) - 1} \qquad (6.58)$$

  with $FSeg \in [\underline{0}, 1]$ defined for $\sharp_P(\mathcal{R}_i) > 1$.
  If we base our considerations on Definition 6.3.3, both characteristics can be quantified using the
- Fragmentation index ($\rho_{Seg.pq}$) (Kang et al., 2009)

$$\rho_{Seg.pq} = \frac{1}{1 + p \cdot |\sharp\mathcal{S} - \sharp\mathcal{N}|^q} \qquad (6.59)$$

  with $\rho_{Seg.qp} \in [0, \underline{1}]$.
- Connectivity coefficient (ρ_{cc}) (Cardenes et al., 2007)

$$\rho_{cc} = \frac{2\min(\sharp\mathcal{A}, \sharp\mathcal{B})}{\sharp\mathcal{A} + \sharp\mathcal{B}} \qquad (6.60)$$

  with $\rho_{cc} \in [0, \underline{1}]$ and symmetric with respect to $\mathcal{A}$ and $\mathcal{B}$ and defined for either $\mathcal{A} \neq \emptyset$ or $\mathcal{B} \neq \emptyset$.

$\rho_{Seg.pq}$ and ρ_{cc} relate to the number of reference segments and not the number of pixels as *FSeg* (Equation 6.58). Furthermore ρ_{cc} can be used locally for each segment using Equation 6.13 and comparing $\mathcal{R}_i$ and $\mathcal{S}_{\mathcal{R}_i}$, for each class comparing $\mathcal{R}_{[k]}$ and $\mathcal{S}_{[k]}$ and the entire sets by $\mathcal{R}$ and $\mathcal{S}$. In the case of using $\rho_{Seg.pq}$ locally for each segment $\mathcal{R}_i$, $\rho_{Seg.11} = 0.5$, if $\mathcal{S}_{R_i} = \emptyset$. The advantage of $\rho_{Seg.pq}$ and ρ_{cc} compared to the approach used in Persello and Bruzzone (2010) is that they only account for the number of assigned segments and not for their geometry. Using these criteria is only meaningful if Equation 6.13 is used for assignment of segments of $\mathcal{S}$ to a reference segment (Sturm and Weidner, 2009; Weidner, 2008).

6.3.4 Combination of Quality Quantities

Most of the quantities discussed above are in the range [0, 1], like values of membership functions within fuzzy classification. Assuming that the best value is 1 for a quantity q, or if not just the quantity $\overline{q} = 1-q$ is considered, in principle the same functions as for the combination and joint evaluation of membership functions can be applied. More generally, functions like the following may be used:

- *Minimum*

$$\rho_{g(min)} = min(\rho_i) \tag{6.61}$$

- *Maximum*

$$\rho_{g(max)} = max(\rho_i) \tag{6.62}$$

- *Mean*

$$\rho_{g(mean)} = \frac{1}{\sum_i w_i} \sum_i w_i \cdot \rho_i \tag{6.63}$$

  used by Heimann et al. (2009) and Persello and Bruzzone (2010)
- *Squared mean*

$$\rho_{g(squared)} = \sqrt{\frac{1}{\sum_i w_i} \sum_i w_i \cdot \rho_i^2} \tag{6.64}$$

  used in Weidner (2008) and by Clinton et al. (2010) with $w_i = 1$.
- *Product*

$$\rho_{g(product)} = \left(\prod_i^n \rho_i \right)^{\frac{1}{n}} \tag{6.65}$$

  used in Sturm and Weidner (2009).

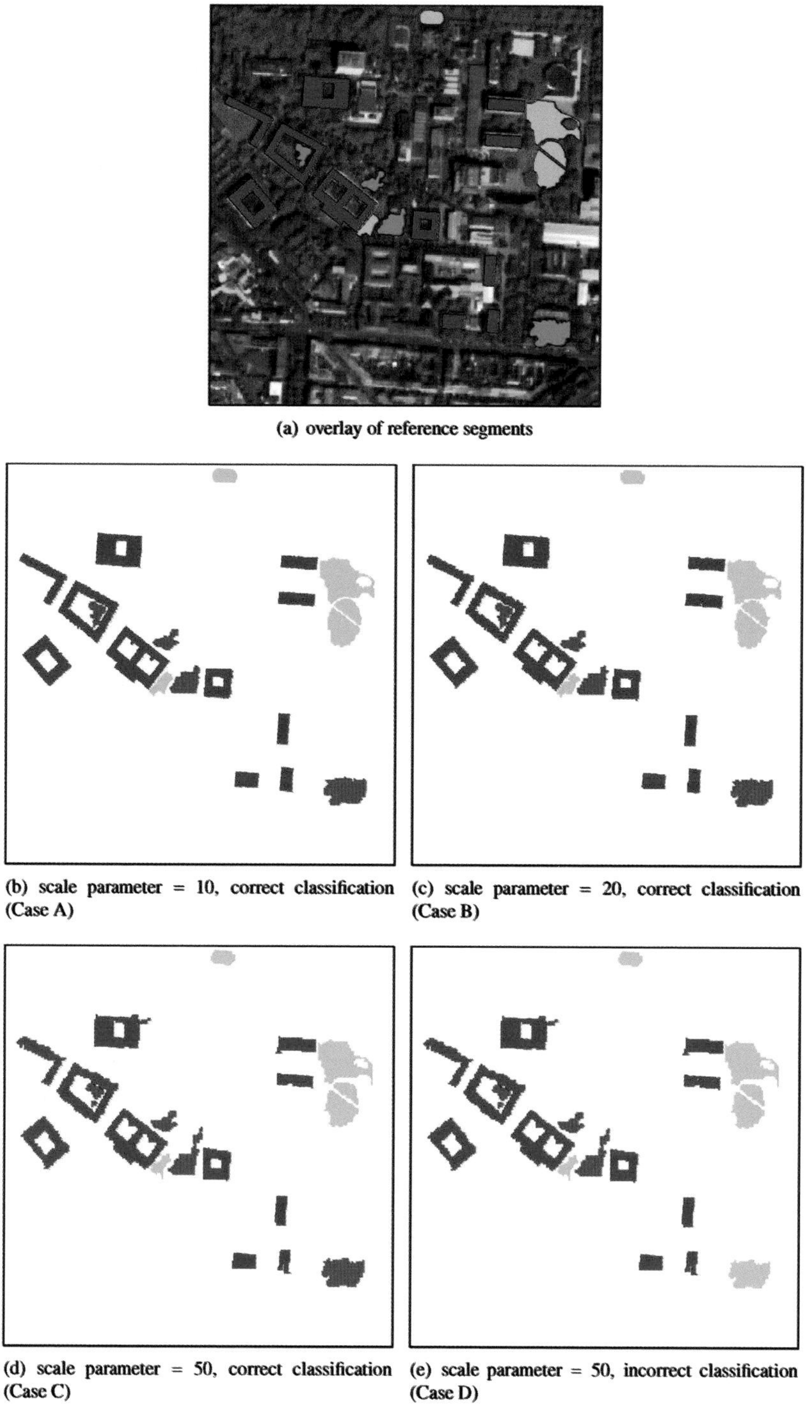

(a) overlay of reference segments

(b) scale parameter = 10, correct classification (Case A)

(c) scale parameter = 20, correct classification (Case B)

(d) scale parameter = 50, correct classification (Case C)

(e) scale parameter = 50, incorrect classification (Case D)

Figure 6.2 Examples.

Table 6.3 Examples – criteria.

	Case A	Case B	Case C	Case D
ρ^*_{OA}	0.997	0.988	0.902	0.886
$OA(\mathcal{R}, \mathcal{S})$	1.000	1.000	1.000	0.947
$\rho_{cc}(\mathcal{R}, \mathcal{S})$	1.000	1.000	0.976	0.976
$\rho^*_{qw}(\mathcal{R}, \mathcal{S})$	0.990	0.968	0.752	0.752
$\rho_{UA[buildings]}$	0.997	0.990	0.933	0.933
$UA(\mathcal{R}_{[buildings]}, \mathcal{S})$	1.000	1.000	1.000	1.000
$\rho_{PA[buildings]}$	0.997	0.990	0.933	0.933
$PA(\mathcal{R}_{[buildings]}, \mathcal{S})$	1.000	1.000	1.000	1.000
$\rho_{cc}(\mathcal{R}_{[buildings]}, \mathcal{S})$	1.000	1.000	1.000	1.000
$\rho^*_{qw}(\mathcal{R}_{[buildings]}, \mathcal{S})$	0.991	0.970	0.813	0.813
$\rho_{UA[trees]}$	0.996	0.990	0.760	0.662
$UA(\mathcal{R}_{[trees]}, \mathcal{S})$	1.000	1.000	1.000	1.000
$\rho_{PA[trees]}$	0.996	0.990	0.760	0.602
$PA(\mathcal{R}_{[trees]}, \mathcal{S})$	1.000	1.000	1.000	0.750
$\rho_{cc}(\mathcal{R}_{[trees]}, \mathcal{S})$	1.000	1.000	0.889	0.857
$\rho^*_{qw}(\mathcal{R}_{[trees]}, \mathcal{S})$	0.989	0.970	0.494	0.339
$\rho_{UA[lawn]}$	0.996	0.988	0.947	0.879
$UA(\mathcal{R}_{[lawn]}, \mathcal{S})$	1.000	1.000	1.000	0.800
$\rho_{PA[lawn]}$	0.996	0.988	0.947	0.947
$PA(\mathcal{R}_{[lawn]}, \mathcal{S})$	1.000	1.000	1.000	1.000
$\rho_{cc}(\mathcal{R}_{[lawn]}, \mathcal{S})$	1.000	1.000	1.000	1.000
$\rho^*_{qw}(\mathcal{R}_{[lawn]}, \mathcal{S})$	0.988	0.964	0.848	0.848

function with a buffer of 1 pixel (Equation 6.69). The influences of segmentation and misclassification for Case D are indicated by the lowest value of ρ_{OA} and the other entries in the upper part of the table. The next entries of Table 6.3 gather the values for $\rho_{UA[k]}$ (Equation 6.72) and $\rho_{PA[k]}$ (Equation 6.73) for the classes *buildings*, *trees* and *lawn* and the values of criteria involved. The results of the class *buildings* are again mainly influenced by the geometric deviations evaluated by the weighted quality rate ρ^*_{qw}, because all segments of class *buildings* are classified correctly. For classes *trees* and *lawn* the results for Case D are also influenced by the misclassification of one *tree* segment as *lawn*.

6.6 Conclusion

In this contribution we have focused on quantitative quality assessment for segmentations and classifications in remote sensing applications. Starting from well known basics like the confusion matrix and quality measures derived therefrom,

Table 6.4 Examples – confusion matrices.

	Case A to C			Case D		
	$\mathcal{R}_{[buildings]}$	$\mathcal{R}_{[trees]}$	$\mathcal{R}_{[lawn]}$	$\mathcal{R}_{[buildings]}$	$\mathcal{R}_{[trees]}$	$\mathcal{R}_{[lawn]}$
$\mathcal{S}_{[buildings]}$	11	0	0	11	0	0
$\mathcal{S}_{[trees]}$	0	4	0	0	3	0
$\mathcal{S}_{[lawn]}$	0	0	4	0	1	4

a number of different criteria also used for quality assessment in computer vision are discussed in detail. These criteria comprise well known criteria for binary classifications, but also criteria that are extended for comparison and evaluation of segmentations. Based on the selected criteria, such as the *overall, user* and *producer accuracy* for the evaluation of classifications per segment and the *connectivity coefficient* and *weighted quality rate*, a concept for quality assessment in remote sensing has been presented. This concept may be adapted from the introduced criteria for computer vision applications by just exchanging one criterion for another that is perhaps more suitable for that specific application.

References

Alt, H., Scharf, L., and Schymura, D. (2008). Probabilistic matching of polygons. In *European Workshop on Computational Geometry (EuroCG), Nancy*, pages 255–258.

Andres, B., Köthe, U., Bonea, A., Nadler, B., and Hamprecht, F. (2009). Quantitative assessment of image segmentation quality by random walk relaxation times. In *Pattern Recognition*, Vol. 5748 of *LNCS*, pages 502–511.

Baatz, M. and Schäpe, A. (1999). Object-oriented and multi-scale image analysis in semantic networks. In *2nd Intl. Symposium. Operationalisation of Remote Sensing, 16–20 August, ITC, NL*, pages 16–20. Herbert Wichmann, Karlsruhe, Germany.

Baddeley, A. (1992). An error metric for binary images. In Förstner, W. and Ruwiedel, S., editors, *Robust Computer Vision*, pages 59–78. Herbert Wichmann, Karlsruhe, Germany.

Baldi, P., Brunak, S., Chauvin, Y., Andersen, C., and Nielsen, H. (2000). Assessing the accuracy of prediction algorithms for classification: An overview. *Bioinformatics Review*, 16(5):412–424.

Baraldi, A., Bruzzone, L., and Blonda, P. (2005). Quality assessment of classification and cluster maps without ground truth knowledge. *IEEE Transactions on Geoscience and Remote Sensing*, 43(4):857–873.

Baumann, A., Boltz, M., Ebling, J., Koenig, M., Loos, H., Merkel, M., Niem, W., Warzelhan, J., and Yu, J. (2008). A review and comparison of measures for automatic video surveillance systems. *EURASIP Journal on Applied Signal Processing*, 2008:1–30.

Boucheron, L., Harvey, N., and Manjunath, B. (2007). A quantitative object-level metric for segmentation performance and its application to cell nuclei. In *International Symposium on Visual Computing*, Vol. 4841 of *LNCS*, pages 208–219.

Cardenes, R., Bach, M., Chi, Y., Marras, I., de Luis, R., Anderson, M., Cashman, P., and Bultuelle, M. (2007). Multimodal evaluation for medical image segmentation. In *CAIP2007*, Vol. 4673 of *LNCS*, pages 229–236.

Chabrier, S., Emile, B., Rosenberger, C., and Laurent, H. (2006). Unsupervised performance evaluation of image segmentation. *EURASIP Journal on Applied Signal Processing*, 2006:1–12.

Cho, K. and Meer, P. (1997). Image segmentation from consensus information. *Computer Vision and Image Understanding*, 68(1):72–89.

Clinton, N., Holt, A., Scarborough, J., Yan, L., and Gong, P. (2010). Accuracy assessment measures for object-based image segmentation goodness. *Photogrammetric Engineering and Remote Sensing*, 76(3):289–299.

Clinton, N., Holt, A., Yan, L., and Gong, P. (2008). An accuracy assessment measure for object based image segmentation. In *IAPRSIS, Vol. 37, Part B4*, pages 1189–1194.

Congalton, R. (2005). Thematic and positional accurracy assessment of digital remotely sensed data. In *7th Annual Forest Inventory and Analysis Symposium*.

Corcoran, P., Winstanley, A., and Mooney, P. (2010). Segmentation performance evaluation for object-based remotely sensed image analysis. *International Journal of Remote Sensing*, 31(3):617–645.

Correia, P. and Pereira, F. (2003). Objective evaluation of video segmentation quality. *IEEE Transactions on Image Processing*, 12(2):186–200.

Crum, W., Camara, O., and Hill, D. (2006). Generalized overlap measures for evaluation and validation in medical image analysis. *IEEE Transactions on Medical Imaging*, 25(11):1451–1461.

Dalirsefat, S., da Silva Meyer, A., and Mirhoseini, S. (2009). Comparison of similarity coefficients used for cluster analysis with amplified fragment length polymorphism markers in the solkworm. *Journal of Insect Science*, 9:1–8.

Estrada, F. and Jepson, A. (2005). Quantitative evaluation of a novel image segmentation algorithm. In *IEEE Computer Society Conference on Computer Vision and Pattern Recognition (CVPR) – Workshops*.

Felzenszwalb, P. and Huttenlocher, D. (1998). Image segmentation using local variation. In *IEEE Computer Society Conference on Computer Vision and Pattern Recognition (CVPR), Washington, DC*.

Filippovska, Y., Kada, M., and Fritsch, D. (2009). Quality evaluation of ground plan generalization. In *12th AGILE International Conference on Geographic Information Science*.

Foody, G. (2001). GIS: The accuracy of spatial data revisited. *Progress in Physical Geography*, 25(3):389–398.

Foody, G. (2008). Harshness in image classification accuracy assessment. *International Journal of Remote Sensing*, 29(10):3137–3158.

Foody, G. (2009). Sample size determination for image classification accuracy assessment and comparison. *International Journal of Remote Sensing*, 30(20):5273–5291.

Förstner, W. (1996). 10 pros and cons against performance characterization of vision algorithms. In *Workshop on Performance Characteristics of Vision Algorithms, Cambridge*.

Fredrich, C. and Feitosa, R. (2008). Automatic adaption of segmentation parameters applied to inhomogeneous objects detection. In *IAPRSIS, Vol. 38, Part 4/C1 – GEOBIA 2008 – Pixels, Objects, Intelligence, Calgary, Alberta, Canada*.

Gaetano, R., Scarpa, G., and Poggi, G. (2009). Hierarchical texture-based segmentation of multiresolution remote-sensing images. *IEEE Transactions on Geoscience and Remote Sensing*, 47(7):2129–2141.

Heimann, T., van Ginneken, B., Styner, M., Arzhaeva, Y., Aurich, V., Bauer, C., Beck, A., Becker, C., Beichel, R., Bekes, G., Belloe, F., Binning, G., Bischof, H., Bornik, A., Cashman, P., Y, C., Cordova, A., Dawant, B., Fidrich, M., Furst, J., Furukawa, D., Grenacher, L., Hornedder, J., Kainmüller, D., Kitney, R., Kobatake, H., Lamcker, H., Lange, T., Lee, J., Lennon, B., Li, R., Li, S., Meinzer, H.-P., Nemeth, G., Raicu, D., Rau, A.-M., van Rixoort, E., Rousson, M., Rusko, L., Saddi, K., Schmidt, G., Seghers, D., Shimizu, A., Slagmolen, P., Sorantin, E., Soza, G., Susomboon, R., Waite, J. M., Wimmer, A., and Wolf, I. (2009). Comparison and evaluation of methods for liver segmentation from CT datasets. *IEEE Transactions on Medical Imaging*, 28(8):1251–1265.

Heipke, C., Mayer, H., Wiedemann, C., and Jamet, O. (1997). Evaluation of automatic road extraction. In *IAPRS, Vol. 32, Part 3-2W3*, pages 47–56.

Henricsson, O. and Baltsavias, E. (1997). 3-D building reconstruction with ARUBA: A qualitative and quantitative evaluation. In Grün, A., Baltsavias, E., and Henricsson, O., editors, *Automatic Extraction of Man-Made Objects from Aerial and Space Images (II)*, pages 65–76. Birkhäuser, Basel, Switzerland.

Hoover, A., Jean-Baptiste, G., Jiang, X., Flynn, P., Bunke, H., Goldgof, D., Bowyer, K., Eggert, D., Fitzgibbon, A., and Fisher, R. (1996). An experimental comparison of range image segmentation algorithms. *IEEE Transactions on Pattern Analysis and Machine Intelligence*, 18(7):673–689.

Jiang, X., Marti, C., Irniger, C., and Bunke, H. (2006). Distance measures for image segmentation evaluation. *EURASIP Journal on Applied Signal Processing*, 2006:1–10.

Kang, W.-X., Yang, Q.-Q., and Liang, R.-P. (2009). The comparative research on image segmentation algorithms. In *International Workshop on Education Technology and Computer Science (ETCS)*.

Karaman, M., Goldmann, L., Yu, D., and Sikora, T. (2005). Comparison of static background segmentation methods. In *Visual Communications and Image Processing (VCIP05), Beijing*.

Keim, D. (1999). Efficient geometry-based similarity search of 3D spatial databases. In *SIGMOD 99, Proceedings ACM SIGMOD International Conference on Management of Data, June 1–3, 1999, Philadelphia, PA*, pages 419–430. ACM Press, New York, USA.

Koch, B., Jochum, M., Ivits, E., and Dees, M. (2003). Pixelbasierte Klassifizierung im Vergleich und zur Ergänzung zum objektbasierten Verfahren. *Photogrammetrie – Fernerkundung – Geoinformation PFG*, 3:195–204.

Kurtz, C., Passat, N., Puissant, A., and Gançarski, P. (2011). Hierarchical segmentation of multiresolution remote sensing images. In *International Smyposium on Mathematical Morphology (ISMM), Intra*, volume 6671 of *LNCS*, pages 343–354.

Looman, J. and Campbell, J. (1960). Adaptation of Sørensen's K (1948) for estimating unit affinities in prairie vegetation. *Ecology*, 41(3):409–416.

Mao, B., Fan, L., Harrie, L., Ban, Y., and Meng, L. (2010). City model generalization quality assessment using nested structures of earth movers's distance. In *13th ICA Workshop on Generalization and Multiple Representation*.

Marcal, A. and Rodrigues, A. (2008). A framework for the evaluation of multi-spectral image segmentation. In *IAPRSIS, Vol. 38, Part 4/C1 – GEOBIA 2008 – Pixels, Objects, Intelligence, Calgary, Alberta, Canada.*

Marpu, P., Neubert, M., Herold, H., and Niemeyer, I. (2010). Enhanced evaluation of image segmentation results. *Journal of Spatial Science*, 55(1):55–68.

Martin, D., Fowlkes, C., Tal, D., and Malik, J. (2001). A database of human segmented natural images and its application to evaluating segmentation algorithms. In *International Conference on Computer Vision (ICCV)*, pages 416–420.

McGlone, J. and Shufelt, J. (1994). Projective and object space geometry for monocular building extraction. In *IEEE Computer Society Conference on Computer Vision and Pattern Recognition (CVPR)*, pages 54–61.

McKeown, D. M., Bulwinkle, T., Cochran, S., Harvey, W., McGlone, C., and Shufelt, J. A. (2000). Performance evaluation for automatic feature extraction. In *IAPRS, Vol. 33, Part B2*, pages 379–394.

Melo, L., Costa, G., Feitosa, R., and Abs da Cruz, A. (2008). Quantum-inspired evolutionary algorithm and differential evolution using in the adaptation of segmentation parameters. In *IAPRSIS, Vol. 38, Part 4/C1 – GEOBIA 2008 – Pixels, Objects, Intelligence, Calgary, Alberta, Canada.*

Neubert, M., Herold, H., and Meinel, G. (2006). Evaluation of remote sensing image segmentation quality–further results and concepts. In *IAPRSIS, Vol. 26, Part 4/C42 – Conference on Object-based Image Analysis (OBIA 2006).*

Niederöst, N. (2001). Automated update of building information in maps using medium-scale imagery. In Baltsavias, E. P., Grün, A., and Gool, L. V., editors, *Automatic Extraction of Man-Made Objects from Aerial and Space Images (III)*, pages 161–170. A. A. Balkema Publishers, Lisse/Abingdon/Exton (PA)/Tokyo.

Ortiz, A. and Oliver, G. (2006). On the use of the overlapping area matrix for image segmentation evaluation: A survey and new performance measures. *Pattern Recognition Letters*, 27:1916–1926.

Paglieroni, D. (2004). Design considerations for image segmentation quality assessment measures. *Pattern Recognition*, 37:1607–1617.

Persello, C. and Bruzzone, L. (2010). A novel protocol for accuracy assessment in classification of very high resolution images. *IEEE Transactions on Geoscience and Remote Sensing*, 48(3):1232–1244.

Pinz, A., Prantl, M., and Ganster, H. (1995). A robust affine matching algorithm using an exponentially decreasing distance function. *Journal of Universal Computer Science*, 1(8):614–631.

Pluempitiwiriyawei, C., Moura, J., Wu, Y.-J., and Ho, C. (2004). Cardiac MR image segmentation: Quality assessment of stacs. In *IEEE International Symposium in BioImaging (ISBI).*

Polak, M., Zhang, H., and Pi, M. (2009). An evaluation metric for image segmentation of multiple objects. *Image and Vision Computing*, 27:1223–1227.

Radoux, J. and Defourny, P. (2006). Influence of image segmentation parameters on positional and spectral quality of the derived objects. In *IAPRSIS, Vol. 26, Part 4/C42 – Conference on Object-based Image Analysis (OBIA 2006).*

Radoux, J., Defourny, P., and Bogaert, P. (2008). Comparison of pixel- and object-based sampling strategies for thematic accuracy assessment. In *IAPRSIS, Vol. 38, Part 4/C1 – GEOBIA 2008 – Pixels, Objects, Intelligence, Calgary, Alberta, Canada.*

Ragia, L., Müller, H., Förstner, W., and Gülch, E. (2000). Qualitätsprüfung von 3D-Stadtmodellen. Technical report, Institut für Photogrammetrie, Universität Bonn, Germany.

Rubner, Y., Puzicha, J., Tomasi, C., and Buhmann, J. (2001). Empirical evaluation of dissimilarity measures for color and texture. *Computer Vision and Image Understanding*, 84:25–43.

Schiewe, J. and Gähler, M. (2006). Modelling uncertainty in high resolution remotely sensed scenes using a fuzzy logic approach. In *IAPRSIS, Vol. 26, Part 4/C42 – Conference on Object-based Image Analysis (OBIA 2006).*

Schiewe, J. and Kinkeldey, C. (2013). Konzeption und Implementierung einer fallspezifischen Unsicherheitskette im Rahmen fernerkundlicher Auswertungen. *Photogrammetrie – Fernerkundung – Geoinformation PFG*, 6:563–574.

Schuster, H.-F. and Weidner, U. (2003). A new approach towards quantitative quality evaluation of 3d building models. In *ISPRS Commission IV Joint Workshop Challenges in Geospatial Analysis, Integration and Visualization II, Stuttgart*, pages 156–163.

Sonka, M. and Fitzpatrick, J., editors (2000). *Handbock of Medical Imaging–Medical Image Processing and Analysis (Vol. 2).* SPIE, Bellingham, USA.

Sørensen, T. (1948). *A Method of Establishing Groups of equal Amplitude in Plant Sociology based on Similarity of Species Content, and its Application to Analyses of the Vegetation on Danish Commons.* Kongelige Danske Videnskabernes Selskabs Biologiske Skrifter.

Srivastava, V. (2006). Evaluation of various segmentation tools for extraction of urban features using high resolution remote sensing data. In *ISPRS TC-IV International Symposium on Geo Spatial Databases for Sustainable Development, Goa, India.*

Stehman, S. and Czaplewski, R. (1998). Design and analysis for thematic map accuracy assessment: Fundamental principles. *Remote Sensing of Environment*, 64:331–344.

Sturm, U. and Weidner, U. (2009). Further investigations on segmentation quality assessment for remote sensing applications. In *ISPRS Workshop High Resolution Earth Imaging for Geospatial Information.*

Udupa, J., LeBlanc, V., Zhuge, Y., Imielinska, C., Schmidt, H., Currie, L., Hirsch, B., and Woodburn, J. (2006). A framework for evaluating image segmentation algorithms. *Computerized Medical Imaging and Graphics*, 20:75–87.

Unnikrishnan, R., Pantofaru, C., and Hebert, M. (2007). Toward objective evaluation of image segmentation algorithms. *IEEE Transactions on Pattern Analysis and Machine Intelligence*, 29:929–944.

Wang, Q., Wang, Q., Mei, X., Zhang, H., and Sun, H. (2009). Quick computation program of fractal dimension for 2D vector data. In *International Conference on Geoinformatics.*

Weidner, U. (2008). Contribution to the assessment of segmentation quality for remote sensing applications. In *IAPRSIS, Vol. 37, Part B7*, pages 1101–1110.

Youden, W. (1950). Index for rating diagnostic tests. *Cancer*, 3:32–35.

Zhan, Q., Molenaar, M., Tempfli, K., and Shi, W. (2005). Quality assessment for geospatial objects derived from remotely sensed data. *International Journal of Remote Sensing*, 26(14):2953–2974.

Zhang, H., Fritts, J., and Goldman, S. (2008). Image segmentation evaluation: A survey of unsupervised methods. *Computer Vision and Image Understanding*, 110:260–280.

Zhang, Y. (1996). A survey on evaluation methods for image segmentation. *Pattern Recognition*, 29(8):1335–1346.

Zhang, Y., Maxwell, T., Tong, H., and Dey, V. (2010). Development of a supervised software tool for automated determination of optimal segmentation parameters for eCognition. In *IAPRSIS, Vol. 38, Part 7B*, pages 287–292.

Zhang, Y.-J. (2001). A review of recent evaluation methods for image segmentation. In *International Symposium on Signal Processing and its Applications (ISSPA), 13–16 August 2001, Kuala Lumpur, Malaysia*, pages 148–151.

Zucker, S. (1976). Region growing: Childhood and adolescence. *Computer Graphics and Image Processing*, 5:382–399.

Part II
Application

From Raw 3D Point Clouds to Semantic Objects

CHAPTER 7

Martin Weinmann, Sven Wursthorn, Boris Jutzi

The fully automatic processing of 3D point clouds represents an issue of major interest in photogrammetry, remote sensing, computer vision and robotics. Nowadays, different subtasks are in the focus of research, including 3D modelling, object extraction and scene interpretation. In this chapter, we address a whole processing pipeline, reaching from the basic task of point cloud acquisition to the high-level task of scene interpretation. Accordingly, point cloud data has to be acquired first which is possible with different techniques (Section 7.1). As a dense and accurate representation of scenes that include objects with possibly homogeneous textures is desirable, we focus on the use of active sensors such as terrestrial laser scanners (TLSs) and range cameras. Subsequently, the acquired point clouds may be analysed with basic point cloud processing techniques. This involves both data representation and point quality assessment (Section 7.2). The results may be further exploited in order to correctly align multiple point clouds in a common coordinate frame, or to enrich given point cloud data with additional information represented as 2D images (Section 7.3). Finally, the representation of larger environments in the form of 3D point clouds and assigned attributes allows a semantic interpretation of the scene (Section 7.4). After explaining the different steps in detail, we provide concluding remarks and suggestions for future work (Section 7.5). In summary, the presented ideas offer a great potential for future research.

7.1 Point Cloud Acquisition

The acquisition of point clouds representing object surfaces and thus 3D shapes has been widely investigated over the last decades. Optical methods, in particular, have gained much importance, as they significantly alleviate a variety of tasks in photogrammetry and computer vision applications, such as the digitisation, reconstruction, interpretation and understanding of static and dynamic 3D scenes. Generally, the different techniques proposed for optical 3D shape acquisition can be categorized according to the measurement principle on which they are based, and numerous valuable surveys can be found in the literature (Dal Mutto et al., 2012; Herbort and Wöhler, 2011; Moons et al., 2010; Sansoni et al., 2009).

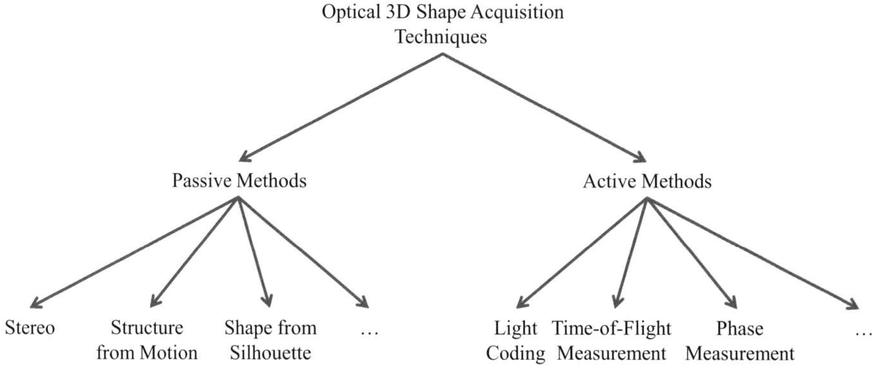

Figure 7.1 Taxonomy of optical 3D shape acquisition techniques.

A common distinction is between passive techniques for scenes with reasonable ambient lighting conditions and active techniques where the scene is actively manipulated, for example by projecting structured light patterns or emitting electro-magnetic radiation in the form of laser light. A further distinction may be introduced that refers to the arrangement of viewpoints and the positions of involved illumination sources. Passive techniques, for instance, can be based on a single view or on multiple views of a scene. Active techniques may rely on a monostatic sensor configuration or a multi-static configuration, where emitter component and receiver component are spatially separated and even several such components may be involved. From these distinctions, a taxonomy of optical 3D shape acquisition techniques can be derived, as illustrated in Figure 7.1. As the scope of this chapter only includes a few categories, only a short description of the most important ideas is provided in the following instead of an exhaustive survey. Depending on the acquisition technique and the used device, the resulting point clouds may be corrupted with noise and, in addition to spatial 3D information, these may also contain respective point attributes such as colour or intensity information. Examples for point clouds acquired via passive and active techniques are illustrated in Figure 7.2.

7.1.1 Passive Techniques

Passive techniques for point cloud generation only rely on 2D image information and thus simple intensity measurements per pixel. For the reconstruction of 3D structures of a scene from such images, different approaches have been proposed of which the most commonly used are considered in the following.

Stereo matching techniques exploit two or more images of a scene, acquired with either multiple cameras or one moving camera, and estimate the respective 3D structure from corresponding points in the images (Brown et al., 2003; Scharstein and Szeliski, 2002). The goal of multi-view stereo (MVS) techniques is to reconstruct a complete 3D object model from a collection of images taken

Figure 7.2 Point cloud derived from 2D images (left) and point cloud derived from direct 3D measurements (right).

from known camera viewpoints (Seitz et al., 2006). The simultaneous recovery of both the 3D structure of a scene and the camera pose from a large number of point tracks (i.e. corresponding image locations) may be obtained with structure-from-motion (SfM) techniques (Szeliski, 2010). A very general approach to SfM that is able to simultaneously recover 3D structures, camera pose and even the intrinsic camera parameters has been proposed with bundle adjustment (Triggs et al., 1999), where bundles of rays connecting the camera centres to 3D scene points are considered and an adjustment is carried out by iteratively minimising the reprojection error.

For several of these approaches, implementations are publicly available. A software system exploiting SfM for structured or unordered image collections is for instance provided with Bundler (Snavely et al., 2006). This system recovers the camera geometry and sparse point clouds representing the local 3D structure as output from a set of images, image features and feature correspondences as input (Snavely et al., 2006). The output can be processed further with a system exploiting MVS such as patch-based multi-view stereo (PMVS) (Furukawa and Ponce, 2010) or clustering views for multi-view stereo (CMVS) (Furukawa et al., 2010) in order to obtain a dense set of patches covering the surface of an object or the 3D structure of a scene. An alternative has been proposed with the combination of APERO as an SfM toolkit and MICMAC as an MVS toolkit (Pierrot-Deseilligny and Cléry, 2011).

7.1.2 Active Techniques

Whereas passive techniques rely on image information arising from simple intensity measurements per pixel, active techniques are based on recording information arising from a signal emitted by the device itself. The emitted signal can either be a coded structured light pattern in the visible or infrared spectrum, or electro-magnetic radiation in the form of modulated laser light.

7.1.2.1 Structured Light Projection

The use of coded structured light can be considered a very reliable technique for recovering object surfaces (Salvi et al., 2004). A device exploiting this technique actively takes an influence on the scene by projecting a coded structured light pattern and thus manipulating the illumination of the scene. Thus, particular labels are assigned to the 3D scene points. When imaging the scene and the projected pattern with a camera, the assigned labels can be decoded in the images. Such synthetically generated features allow for establishing robust feature correspondences, and the respective 3D coordinates can easily be recovered by triangulation. Techniques based on the use of structured light patterns can be categorised depending on the pattern codification strategy (Dal Mutto et al., 2012; Salvi et al., 2004):

- Direct codification strategies exploit a pattern that is typically based on a large range of either grey or colour values. Thus, each pixel can be labelled with the respective grey or colour value of the pattern at this pixel.
- Time-multiplexing strategies exploit a temporal coding, where a sequence of structured light patterns is projected onto the scene. Thus, each pixel is assigned a codeword consisting of its illumination value across the projected patterns. The respective patterns may, for instance, be based on binary codes or Gray codes and phase shifting.
- Spatial neighbourhood codification strategies exploit a unique pattern. The label associated to a pixel is derived from the spatial pattern distribution within its local neighbourhood. Thus, labels of neighbouring pixels share information and provide an interdependent coding.

7.1.2.2 Time-of-flight (ToF)/Phase Measurements

A different technology for measuring the distance to object surfaces is provided by systems that illuminate a scene with modulated laser light and analyse the backscattered signal. Such techniques based on emitting electro-magnetic radiation can be categorised according to the laser type, the modulation technique, the measurement principle, the detection technique or the arrangement of emitting and receiving components of the device (Stilla and Jutzi, 2009). In general, a laser pulse is emitted by the device and transmitted to an object. At the object surface, the pulse is partially reflected and, finally, a certain amount of the original laser pulse reaches the receiver unit of the device. The measurement principle is therefore of great importance as it may be based on different signal properties such as amplitude, frequency, polarisation, time or phase.

The time t between emitting and receiving a laser pulse corresponds to the respective time-of-flight, and hence this principle is used by devices commonly referred to as time-of-flight (ToF) cameras. For each pixel, the distance r to the

the perspective-n-point (PnP) problem. In general, solving the PnP problem, also known as the exterior orientation problem, means estimating a camera pose from a set of n known 3D points and their observed 2D image locations (Fischler and Bolles, 1981; Wu and Hu, 2006).

The transfer to point cloud registration requires 2D observations in the image plane, which is available when using a range camera such as Microsoft Kinect, PMD[vision] CamCube 2.0 or MESA Imaging SR4000 (Weinmann et al., 2011b; Weinmann and Jutzi, 2012). In the case of terrestrial laser scans where the measured intensity values are sampled on a spherical grid, a camera-like projection onto a virtual plane has to be introduced (Weinmann et al., 2011a). Generally, the position and orientation of the scanner are assumed to be known for one scan (e.g. scan $\mathcal{S}_k$) with respect to a global coordinate frame, and therefore the respective 3D coordinates $\mathbf{X}_{i,k}$ of the forward projected 2D features can easily be derived with respect to the global coordinate frame. Consequently, the camera-like projection has to be introduced for scan $\mathcal{S}_{k+1}$. The respective projection model represents a pinhole-camera model for backprojecting the respective 3D coordinates $\mathbf{X}_{i,k+1}$ known in the local coordinate frame according to

$$\mathbf{x}_{i,k+1} = \mathbf{K}_{\mathcal{C}} \left[\mathbf{R}_{\mathcal{C}} | \mathbf{t}_{\mathcal{C}}\right] \mathbf{X}_{i,k+1} \tag{7.8}$$

onto subpixel-accurate 2D points $\mathbf{x}_{i,k+1}$ on the image plane of a virtual camera $\mathcal{C}$, where $\mathbf{K}_{\mathcal{C}} \in \mathbb{R}^{3\times3}$ is the respective calibration matrix (Figure 7.6). The matrix $\mathbf{R}_{\mathcal{C}} \in \mathbb{R}^{3\times3}$ describes the orientation of the virtual camera with respect to the local coordinate frame, and the vector $\mathbf{t}_{\mathcal{C}} \in \mathbb{R}^3$ represents the respective translation between the position of the scanner and the projective centre of the virtual camera.

Since a virtual camera is involved, we may specify its pose and its interior behaviour without loss of generality. Accordingly, $\mathbf{R}_{\mathcal{C}}$ and $\mathbf{t}_{\mathcal{C}}$ may be defined so that the virtual camera looks into the horizontal direction and its projective centre coincides with the position of the scanner (Weinmann et al., 2011a). Furthermore, any values can be chosen to describe its intrinsic parameters, and the image plane does not even have to be limited on a finite area, as only the 2D projections onto the introduced image plane of the virtual camera are of interest. We may further assume that the relative orientation of the virtual camera $\mathcal{C}$ remains fixed with respect to the local coordinate frame of the sensor for the considered scans $\mathcal{S}_k$ and $\mathcal{S}_{k+1}$.

The derived 3D/2D correspondences $\mathbf{X}_{i,k} \leftrightarrow \mathbf{x}_{i,k+1}$ are used for solving the PnP problem, where the efficient-perspective-n-point (EPnP) algorithm (Moreno-Noguer et al., 2007; Lepetit et al., 2009) provides a robust and efficient solution with only linear complexity. For increased robustness, a RANSAC-based EPnP scheme is used (Weinmann et al., 2011a). Optionally, the influence of each 3D/2D correspondence on the registration process can be weighted, for example with a weighting scheme derived from the reliability of the respective 3D points in the form of inverse cumulative histograms (Weinmann and Jutzi, 2013).

information about radiometric and geometric surface properties of observed objects. This might not always be sufficient. For a more detailed interpretation of a scene, additional attributes might be helpful. In this regard, colour information captured with a digital camera can be used to classify the material of an object surface or to detect vegetation, whereas thermal information captured with a thermal camera represents an attribute that offers a different look on the respective objects and even on their interior behaviour. In particular, the attribute of temperature is important for building inspection (Hoegner et al., 2007; Iwaszczuk et al., 2011). Different materials, concealed objects such as heating pipes and structural defects in the form of temperature leakages can easily be identified and added as semantic or geometric information to a building model. Furthermore, temperature allows for easily detecting people within the observed scene (Weinmann et al., 2012). Hence, a sensor platform equipped with different devices offers great advantages. However, such a platform requires the acquisition of co-registered information, which means that the relative orientation between the different sensors has to be determined initially. Once the relative orientation is known, the relations between complementary measurements can easily be derived.

In general, devices such as terrestrial laser scanners, range cameras, RGB cameras, thermal cameras and multi-spectral sensors allow for image representations of the measured information. Hence, the relative orientation between the different devices can also be estimated only implicitly by directly estimating a transformation between the respective images. In order to estimate this transformation, corresponding information has to be detected within these images. In this context, however, the detection of wavelength-independent properties to derive correspondences between the different spectral domains is a major challenge. Objects visible in the visual domain may for instance be hardly visible or even invisible in the thermal infrared domain if they have the same temperature as the respective background, whereas in thermal infrared images, objects may be visible that are below the surface of another object and thus remain invisible in the visual domain. Furthermore, contours of objects appear blurred in thermal infrared images and do not show strong edges, which are usually present in intensity or colour images. The attribute of colour might reveal additional information compared to intensity measurements, as the conversion between grey-valued intensity images and colour images contains ambiguous mappings, i.e. several colour values might be represented with the same intensity value.

Whereas a fully automatic detection of point correspondences via standard techniques such as the SIFT algorithm (Lowe, 2004) typically fails due to the different characteristics of the compared images, the shape-based matching algorithm (Steger, 2001; Ulrich, 2003) is still able to derive feature correspondences, of which a certain percentage provide high reliability. In general, the shape-based matching algorithm consists of several steps. First, a model image has to be generated, which is achieved, for instance, by selecting quadratic areas of 100 × 100 pixels around points of a discrete grid with a grid spacing of 10 pixels in the first image. Subsequently, a gradient filter is applied to the model image, and the associated gradient directions are determined. The model image is matched to the gradients of the search image by comparing the respective gradient directions. In particular, a similarity measure is calculated, which represents the average vector product of the gradient directions of the transformed model and the search image. More details are described in (Weinmann et al., 2012).

Once corresponding information has been detected, the straightforward approach is to exploit this information for adequately mapping one of the image representations onto the other one. This alignment is commonly referred to as image registration. Valuable surveys of different approaches for image registration are provided in (Zitová and Flusser, 2003) and (Szeliski, 2006). Common approaches rely on point features and estimate a 2D or 3D transformation between sets of matched 2D or 3D points (Szeliski, 2010). This involves either a parametric transformation (i.e. a rigid transformation), mapping features in one image to corresponding features in another image, or a non-rigid transformation that also allows for elastic deformations. Considering a displacement between two sensors and also a possible difference with respect to the sensor orientation, the respective transformation has to satisfy certain constraints. Among the parametric transformations, a projective or perspective transformation can for instance be applied to adequately describe the relation between the respective images in the case of planar scenes. For such a planar scene, the images are related with a homography, where 2D points of one image plane are mapped onto 2D points of the other image plane (Brown and Lowe, 2003; Hartley and Zisserman, 2008). For non-planar scenes, the relation between two images can be described with a fundamental matrix encapsulating the intrinsic projective geometry between two views (Barreto and Daniilidis, 2005; Hartley and Zisserman, 2008). Following (Weinmann et al., 2012), we apply shape-based matching, assume a planar scene and exploit the transformation model of a homography. The resulting infrared textured 3D models are shown in Figure 7.8.

Figure 7.8 3D point clouds with co-registered thermal infrared information for an almost planar scene (left) and a non-planar scene (right).

7.4 Scene Analysis

Once a scene is adequately represented as a 3D point cloud, a challenging task is to automatically interpret the given point cloud. This implies that each 3D scene point should be labelled with a semantic label (e.g. *ground, building* or *vegetation*). Given a large and densely sampled environment captured with a terrestrial or mobile laser scanner, scene analysis may be conducted by exploiting geometric information, which is implicitly represented by the spatial arrangement of 3D points. In order to derive an adequate description of this implicit information, the common strategy is to first recover the local neighbourhood of each 3D point and subsequently calculate various geometric features (Section 7.4.1). These are for instance 3D features that describe properties of the local object surface around a certain 3D point. Further 2D features can be derived from the projection of the 3D point cloud onto a (e.g. horizontally oriented) 2D plane (Weinmann et al., 2013). Using adequate training data, the whole feature set or only a subset may be used in order to classify 3D points according to the local surface structure they describe (Section 7.4.2). Thus it can be distinguished if a 3D point describes 1D structures such as wires, poles and trunks or 2D structures such as walls, doors and building façades or 3D structures arising from vegetation such as foliage and objects without a specific shape (Demantké et al., 2011; Weinmann et al., 2013).

Here, we focus on the consideration of individual 3D points, which does not include contextual information. Learning such contextual relationships would require more complex models such as Markov random fields (MRFs) or conditional random fields (CRFs) and thus cause more computational effort. Instead, we use standard classifiers and try to improve the classification accuracy by only selecting the most relevant features from the whole feature set (Section 7.4.3). Such a feature selection also reduces both computational effort and memory consumption.

7.4.1 Feature Extraction

For feature extraction, we apply the definition according to which a local 3D neighbourhood is formed by a point and its k closest neighbours (cf. Part I, Chapter 3). From the spatial 3D coordinates of these points, the 3D structure tensor is derived. Its eigenvalues are exploited to derive the local 3D shape features represented by linearity L_λ, planarity P_λ, scattering S_λ, omnivariance O_λ, anisotropy A_λ, eigenentropy E_λ, change of curvature C_λ and the sum of eigenvalues denoted as Σ_λ. Furthermore, the height Z, the verticality V, the local point density D and the radius $r_{k\text{-NN}}$ of the neighbourhood are used, as well as the height difference $\Delta Z_{k\text{-NN}}$ and the height variance $\sigma_{Z,k\text{-NN}}$ within the local neighbourhood. This yields a set consisting of 14 derived 3D features.

As described in Part I, Chapter 3, we further derive 2D features by projecting all the 3D points onto a horizontally oriented plane. This yields features in the form of the local point density D_{2D} and the radius $r_{k\text{-NN},2D}$ for the considered neighbourhoods in the new 2D representation. Furthermore, the sum $\Sigma_{\lambda,2D}$ and the ratio $R_{\lambda,2D}$ of the eigenvalues of the respective 2D structure tensor are used as 2D features. Finally, a quantisation into quadratic bins of side length 0.25 m yields a discrete accumulation map, and the number M of points falling into the same bin as the considered 3D point is defined as a 2D feature as well as the height difference ΔZ and the height variance σ_Z of all points within the respective bin. This yields a set consisting of seven derived 2D features.

7.4.2 Semantic Scene Interpretation

Given an at least partially labelled 3D point cloud as ground truth data, it is possible to train a classifier in order to adequately label the remaining part of the point cloud or to generalise towards similar 3D scenes. For this purpose, a training set $\mathcal{X} = \{s_i\}$ is required that consists of n single training samples s_i with $i = 1, \ldots, n$. Each training sample $s_i = (\mathbf{x}_i, l_i)$ encapsulates a feature vector $\mathbf{x}_i \in \mathbb{R}^d$ in a d-dimensional feature space and the respective class label $l_i \in \{1, \ldots, n_C\}$, where n_C represents the number of classes. The test is performed on a disjoint test set $\mathcal{Y} = \{s_j\}$. This test set consists of m single samples s_j with $j = 1, \ldots, m$, which encapsulate new feature vectors $\mathbf{x}_j \in \mathbb{R}^d$ in the d-dimensional feature space and, optionally, for evaluating the classification results, the respective class labels $l_j \in \{1, \ldots, n_C\}$.

For evaluation, we use the Oakland 3-D Point Cloud Dataset[2] (Munoz et al., 2008, 2009a). This benchmark data set was acquired in an urban environment with a mobile platform equipped with side-looking SICK LMS laser scanners used in push-broom mode. The whole data set is separated into three disjoint subsets representing a training set $\mathcal{X}$, a validation set $\mathcal{V}$, and a test set $\mathcal{Y}$. Each 3D point of the subsets is labelled with one of the five semantic labels *wire*, *pole/trunk*, *façade*, *ground* and *vegetation*.

Instead of complex learning approaches involving contextual information (Munoz et al., 2009a,b; Niemeyer et al., 2012; Shapovalov and Velizhev, 2011), we follow the strategy of using standard classifiers in the form of a nearest neighbour (NN) classifier, a k nearest neighbour (k-NN) classifier with $k = 50$, a naive Bayesian (NB) classifier and a multi-class support vector machine (SVM) classifier with a Gaussian kernel. For each 3D point, we calculate the whole set of 3D and 2D features by considering the spatial arrangement of this point and its $k = 50$ closest neighbours. The feature vectors are used to train the classifiers on the training set $\mathcal{X}$. As the number of samples per class varies significantly for both training set $\mathcal{X}$ and validation set $\mathcal{V}$, we reduce these sets to equal class size by selecting all samples of the smallest class and randomly selecting the same number of samples for all other classes. The training process is performed on the reduced training set $\mathcal{X}_r$ and, for the SVM classifier, validation is carried out with the reduced validation set $\mathcal{V}_r$ in order to find the optimal choice for the width γ of the Gaussian kernel and the parameter C penalising classification errors based on a grid search in a suitable subspace (γ, C). For all classifiers, a subsequent

Table 7.1 Confusion matrix resulting from SVM-based classification involving all 3D and 2D features. Reference labels are provided in the left column, and estimated labels are represented in the upper row. The overall classification accuracy is 89.48%.

Class	Wire	Pole/Trunk	Façade	Ground	Vegetation	Recall
Wire	2414	538	300	41	501	0.636
Pole/Trunk	1552	4691	219	41	1430	0.591
Façade	10,480	33,298	63,335	62	3937	0.570
Ground	12,477	2785	223	914,461	4200	0.979
Vegetation	24,027	24,349	13,600	5229	200,120	0.749
Precision	0.047	0.071	0.815	0.994	0.952	

performance evaluation is carried out on the full test set $\mathcal{Y}$. Exemplary results for SVM-based classification involving all 3D and 2D features are provided in Table 7.1.

7.4.3 Feature Relevance Assessment

Concerning feature extraction, it has to be considered that the derived features may contain redundant or irrelevant information with respect to the classification task. Hence, it would be desirable to select only a compact subset of the most relevant features. This may increase the classification accuracy, and reduce computational effort with respect to processing time and memory consumption due to much less involved data. Recent approaches for feature selection can be categorised into filter-based methods, wrapper-based methods or embedded methods (Guyon and Elisseeff, 2003). A filter-based selection is based on a performance evaluation function, providing a score that can be directly calculated from the training data without involving learning techniques. Such a function is typically derived from a common measure for distance, information, dependency or consistency, which means that it evaluates intrinsic properties of the given training data. In contrast, a wrapper-based selection directly involves the performance of a classifier in order to evaluate each subset and thus conduct feature selection. An embedded method is based on searching the parameter space of a classifier and the feature selection space simultaneously, and thus feature selection is carried out during the training process.

In order to derive compact and robust classifier-independent subsets of versatile features with general applicability, we analyse the relevance of the derived features according to a general relevance metric addressing different intrinsic properties of the given training set (Weinmann et al., 2013). This metric takes into account the mean rank across seven different relevance measures. Thus, the global rank of each feature is derived according to the ascending order of the mean rank across all seven measures. Relevant features can thus be selected

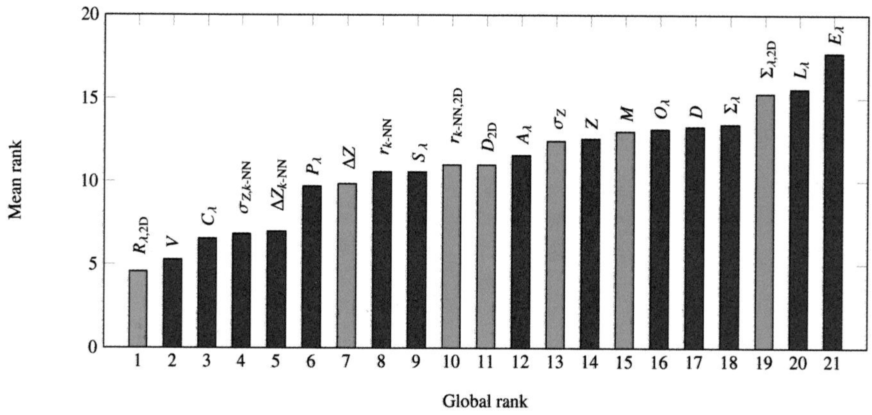

Figure 7.9 Mean rank of the derived 3D features (blue) and those features derived from considerations in 2D (green).

by only keeping a feature subset consisting of the best-ranked features. For the aforementioned example of semantically interpreting 3D point cloud data, the ranking with respect to the combined relevance metric is illustrated in Figure 7.9.

The combination of different relevance measures for feature ranking reveals that the ratio $R_{\lambda,2D}$ of the eigenvalues in 2D, the verticality V and the change of curvature C_λ are highly relevant. The feature $R_{\lambda,2D}$ is a strong feature for detecting building façades. The verticality V (which is based on the Z-component of the local normal vector) is well-suited to distinguish between façades and ground, and the change of curvature C_λ is useful for distinguishing between planar structures such as façades or ground, and non-planar structures representing vegetation. Furthermore, the height difference $\Delta Z_{k\text{-NN}}$ as well as the height variance $\sigma_{Z,k\text{-NN}}$ within the local neighbourhood are considered to be highly relevant. In contrast, the features Σ_λ, $\Sigma_{\lambda,2D}$ and E_λ are among the least relevant features. This is consistent with other investigations involving echo-based and full-waveform laser scanning features (Mallet et al., 2011). Also the linearity L_λ does not seem to be a relevant feature in the experiment (Figure 7.9). This is caused by the fact that a line-of-sight instrument is used for data acquisition, and so high values mainly arise at edges of objects where a linear behaviour is indicated, although there is no linear structure. Furthermore, we can state that there are more and less relevant features among both 3D features and 2D features.

Exploiting the principle of forward selection, we successively train and test the classifiers with one additional feature per iteration, where the derived order of the features is used. Between the features with global rank 5 and global rank 6, the derived relevance metric shows a significant change (Figure 7.9). For this reason, we assume the five best-ranked features to be a compact and robust subset of versatile features for the experiments. Following the classification strategy described in Section 7.4.2, the performance of different classifiers is tested on

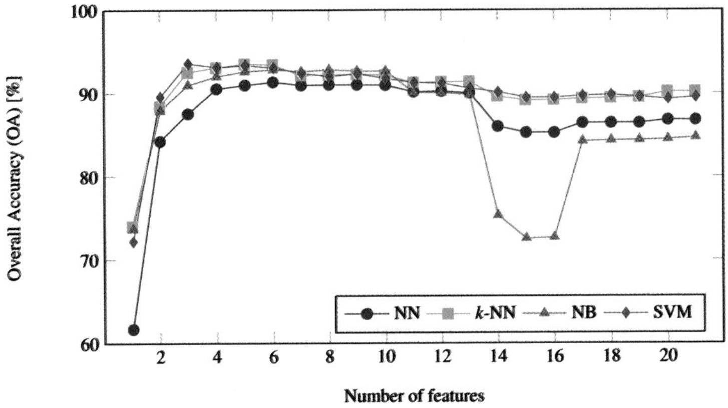

Figure 7.10 Overall classification accuracies for different classifiers. The size of the feature subset is iteratively increased according to the ascending global rank depicted in Figure 7.9.

Table 7.2 Confusion matrix resulting from SVM-based classification involving only the five best-ranked features. The overall classification accuracy is 93.32%.

Class	Wire	Pole/Trunk	Façade	Ground	Vegetation	Recall
Wire	1831	125	724	208	906	0.483
Pole/Trunk	1378	4729	386	8	1432	0.596
Façade	7506	5577	88,664	237	9128	0.798
Ground	4680	501	191	917,418	11,356	0.982
Vegetation	19,042	10,308	11,277	3,550	223,148	0.835
Precision	0.053	0.223	0.876	0.996	0.907	

various feature subsets. The resulting overall accuracies are depicted in Figure 7.10 and clearly reveal that the highest classification accuracies are obtained for relatively small feature subsets. Furthermore, it becomes clear that including the height Z with global rank 14 decreases the overall classification accuracy, in particular for the NB classifier, which might be due to the fact that elements of the different classes appear at different heights. For the subset of the five best-ranked features, the respective confusion matrix for SVM-based classification is provided in Table 7.2, and a visual impression of the respective classification results is provided in Figure 7.11. This feature subset even yields precision/recall values that are comparable to recent investigations, where associative Markov networks and thus graphical models involving contextual information have been used (Munoz et al., 2009b).

The resulting overall classification accuracies for other commonly used feature subsets are provided in Table 7.3. Accordingly, the classification based on the five best-ranked features significantly outperforms classification using the

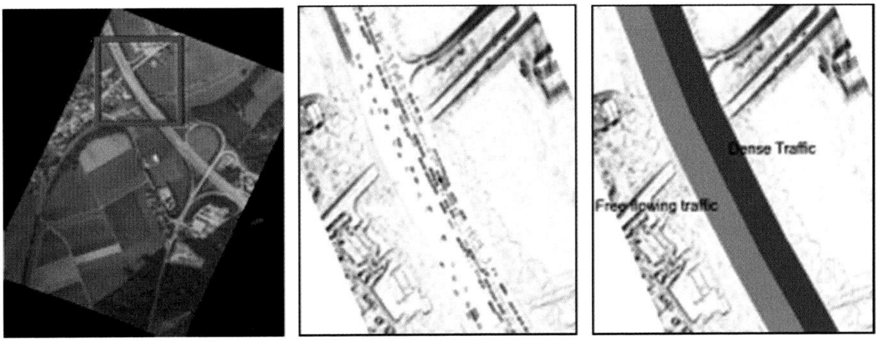

Figure 8.5 Left: Geocoded image. Centre: Temporal grey variance. Right: traffic type classification in RoIs.

texture measures such as local spectral variance and granularity. More features are introduced in case stereo pairs (e.g. IKONOS single-pass stereo images) or airborne image sequences have been recorded. These allow us to exploit time-dependent object attributes as defined in the traffic model, in particular the local grey value variance measured along the time axis. After classification morphological cleaning is applied to separate road regions in RoIs containing free-flowing traffic from those containing dense traffic (Figure 8.5).

8.3.3 Analysis of Free-Flowing Traffic

The analysis of free-flowing traffic starts with the segmentation of paved regions using intensity-hue-saturation (IHS) colour space. This colour space is particularly suitable because pavement pixels generally show low colour saturation. Even if hue and intensity may vary, road pixels nevertheless form dense clusters around the medial axis of the IHS cone. As a second feature for segmentation, the average spectral distance between a pixel's colour and the colour of its local neighbourhood is included. While the first feature captures the global spectral properties of roads, the latter emphasises the typical local road homogeneity.

The modules called during the analysis of free flowing depend on the respective input data. If only monocular images are available (which is a typical situation for spaceborne images), the extraction of cars must purely rely on geometric and radiometric characteristics. Vehicles are simply modelled as compact, coloured image regions ("blobs"), since detailed sub-structures of cars disappear almost completely at a resolution of 0.6 m and coarser. The car blobs are extracted in all non-road regions of the RoIs through a differential-geometric blob detector applied to the panchromatic channel; see Hinz (2005) for details. After initial extraction, only those blobs are kept whose dimensions fit cars or medium-sized trucks. In addition, the spectral signature of a blob's interior must form a dense cluster in hue-saturation space; that is, the second-order central moments of the cluster must be small. Figure 8.6 shows the result

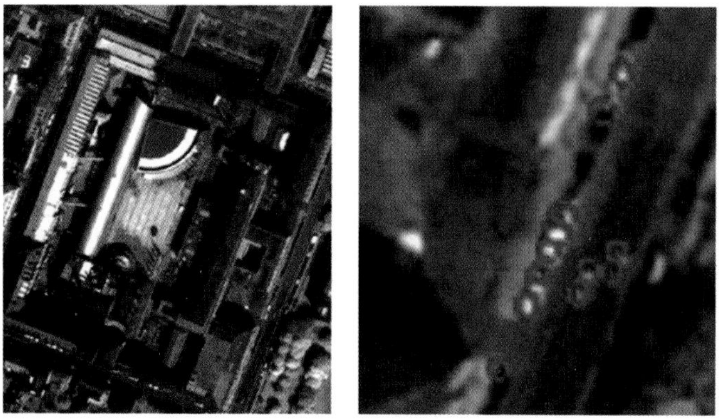

Figure 8.6 Left: Classified road pavement for roads selected from GIS. Right: Detected blobs in RoIs.

of road pavement extraction and various detected cars in a cut-out of a QuickBird scene. The next steps are devoted to the estimation of vehicle velocities and depend on the available input data.

8.3.3.1 Velocity Estimation from Satellite Images

To estimate movements from monocular satellite images, one can take advantage of the fact that the optical axes of the panchromatic and multi-spectral channels of spaceborne systems such as IKONOS or Quickbird do not coincide. The time gap that appears between the acquisition of the panchromatic and the multi-spectral data can be used to derive velocity information. To estimate the translation vector between the blob of the input region from the pan channel and its conjugate region in the multi-spectral data, a shape-based matching approach based on the work of Steger (2001) is employed. The matching result is further refined by least-squares fitting of Gaussian kernels to the grey values of the found blob regions of the pan and multi-spectral channels, respectively. The incorporation of the least-squares estimation provides the basis to reason about the accuracy of the movement estimates and to apply statistical testing to decide whether an object moves at all. As derived in Leitloff et al. (2007b, 2010), the lower bound of detectable velocities in QuickBird images is around 20 km/h. Figure 8.7 illustrates the initial detection in the pan channel (green), the matched point in the multi-spectral channels (red) and the fitted Gaussian kernels to estimate the final positions for motion determination from monocular satellite images.

If more than one image is available, additional motion features can be exploited. For the case of satellite single-pass stereo images, the time difference between the two acquisitions is unfortunately too large to apply a tracking approach (typically 6–12 s between acquisitions). A straightforward approach to confirm a moving car hypothesised by the above scheme would consist of simply checking whether the extracted blob boundary falls into a road region

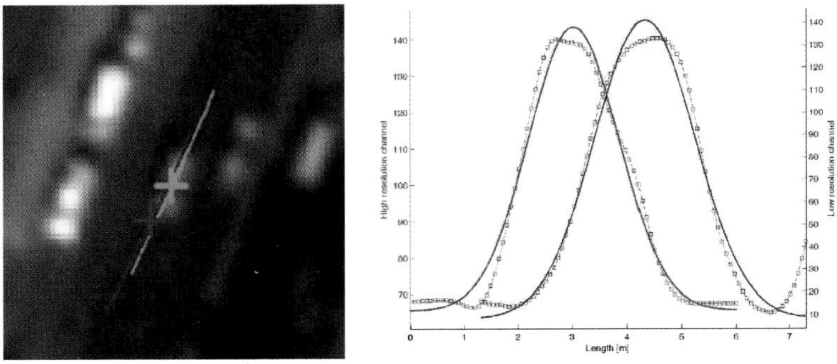

Figure 8.7 Left: Initial detection in pan channel (green), matched position in multi-spectral data (red) and profiles for fitting Gaussian kernels. Right: Fitted Gaussian kernels for estimating final motion vector.

of the next image. Such an approach might, however, lead to numerous missing detections due to erroneous road classification results. A better alternative is therefore the extension of the 2D hue-saturation feature space with an additional dimension capturing the spectral distances of colour values of the blob interior between the first and second images. This distance should be large for moving cars because, once the car has moved after the first image acquisition, it is very unlikely that this region shows similar spectral characteristics in the second image.

8.3.3.2 Velocity Estimation from Airborne Images

In the airborne case, highly overlapping image sequences are usually recorded. While video cameras allow for almost continuous image acquisition, the frame rates of airborne cameras typically range from $\frac{1}{3}$ Hz to 3 Hz or even more. In the following, the focus is on airborne frame cameras such as the 3 K camera described in Thomas et al. (2009) and Kurz et al. (2012). They have a much wider field-of-view than video cameras, which makes them more suitable for capturing the traffic situation of a large road network.

Initial vehicle candidates are extracted with the approach for monocular images described above. Image triplets are then used in each vehicle tracking step in order to gain a certain redundancy, allowing an internal evaluation of the results. The initial state of a car is described by two parameters, the distance to the road side polygon and the approximate motion direction. Then, a vehicle image model is created by selecting a rectangle around the car (no. (1) in Figure 8.8). By using again the shape-based matching algorithm (Steger, 2001), car hypotheses are found in the following images. In order to reduce the search space for the matching procedure, a RoI is derived from the motion-dependent attributes in the traffic model. The matching procedure delivers matches M12 in Image 2 and matches M13 in Image 3 (2). Both M12 and M13 may contain multiple match results, also including some wrong matches (Figure 8.8). For

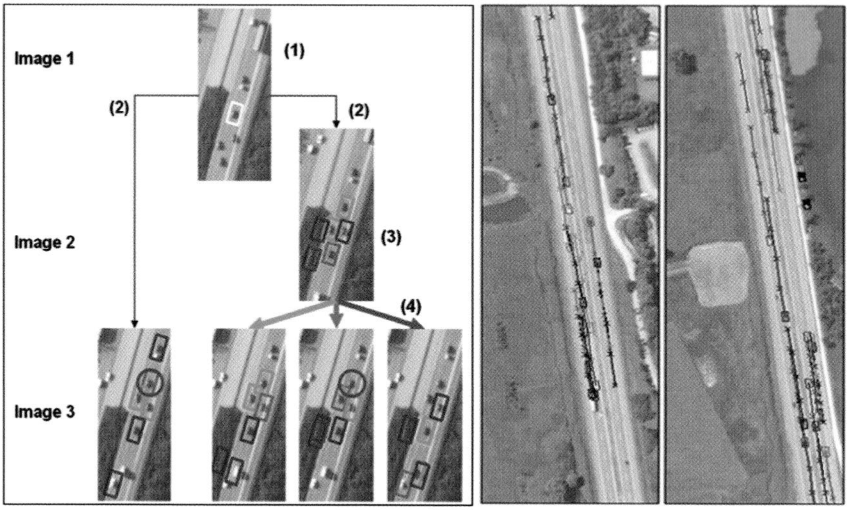

Figure 8.8 Left: Multiple hypotheses tracking over image triplets. Blue circles indicate correct tracks. Right: Two examples of tracked vehicles over a sequence of five images. Colours indicate different tracks.

each match M12, vehicle state parameters are calculated and new vehicle image models are created based on the match positions of M12 (3). These models are searched in Image 3 (4), eventually resulting in matches M23, for which again vehicle parameters are determined. Finally, the results M13 and M23 are checked for consistency including a smoothness criterion of the trajectory to determine the correct combination of the matches (marked as blue circle). Figure 8.8 (right) visualises two examples of tracked cars over a sequence of five images. Please note the sparse temporal sampling of the sequence, which is indicated by the large translation vectors in the tracks. A detailed description of the tracking approach can be found in Lenhart and Hinz (2006).

8.3.4 Analysis of Dense Traffic

In the case of dense traffic the order of the cars inside a row hardly changes from one image to the other (cf. the traffic model). It is therefore not necessary to track cars separately from each other to obtain an estimate of the current velocity. Instead, the local velocity at a certain road section can be estimated by calculating local shifts between consecutive image pairs. It is nonetheless advisable to mask static regions to avoid biased velocity estimates. Hence, all non-textured regions are excluded before applying the shift estimation algorithm. All road RoIs are straightened by resampling the images parallel to the GIS- axes, so that an efficient algorithm can be applied, as originally developed for disparity estimation in epipolar images. Please note that, in contrast to the tracking approach described above, local shift estimation enforces constant topology of

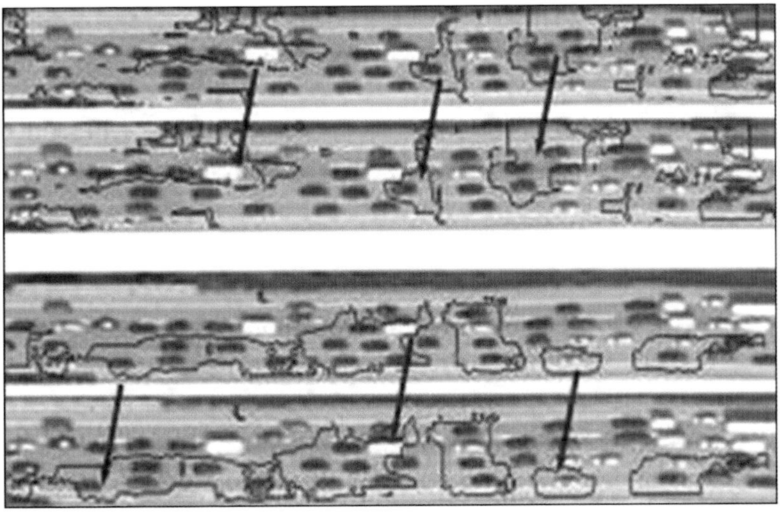

Figure 8.9 Segmentation examples based on local shifts. Top: Regions of five pixels shift projected into the image pair. Bottom: Regions of six pixels shift projected into the image pair.

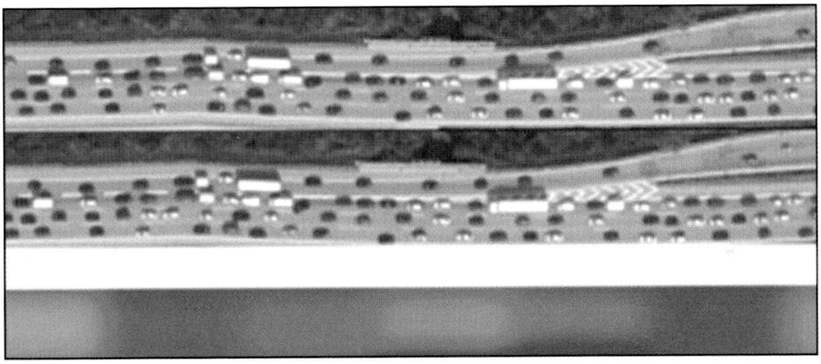

Figure 8.10 Top: Image pair straightened along the GIS axis. Bottom: Velocity pattern for the corresponding highway section derived from the local shift map. Brighter values indicate faster motion.

the objects on ground, since the lines of an image pair are transformed onto each other without changing the pixel order.

The application of this algorithm to one image pair is shown in Figure 8.9. This visualises two examples when thresholding the shifts and projecting the resulting regions back into the image pair. The complete velocity information derived from local shifts is shown in Figure 8.10. This example clearly indicates the well-known periodic velocity pattern of traffic congestion. Where the velocities are different for individual lanes of a road, one can optionally infer the lanes from the velocity profiles.

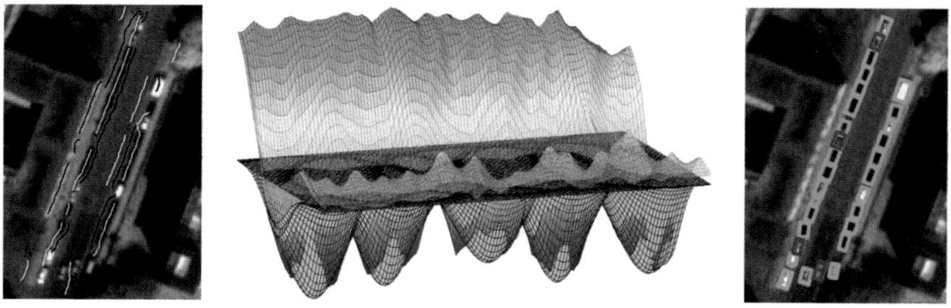

Figure 8.11 Left: Extracted ribbons (dark queues, white; bright queues, black). Centre: Fitting of Gaussian kernels to grey value profiles. Right: Vehicles reconstructed from estimated values (correct detections, green; missing detections, red).1.11 Left: Extracted ribbons; dark queues (white), bright queues (black). Centre: Fitting of Gaussian kernels to gray value profiles. Right: Vehicles reconstructed from estimated values; correct detections (green), missing detections (red).

8.3.5 Detection of Vehicles in Dense Traffic

The analysis of dense traffic yields image regions that are static or slowly moving, yet without any semantics. Individual vehicles have not been extracted up to now. Hence, the fourth unit aims at reconstructing the individual vehicles for both parking and slowly moving vehicle lines (cf. traffic model); that is, those regions with local shift close to zero (i.e. no movements) are processed separately from the regions with larger shifts. A detailed description of the extraction of vehicles parked in rows can be found in Leitloff et al. (2010) and Leitloff (2011). In the following, only the fundamental steps are summarised.

The procedure is based on a "coarse-to-fine" strategy; that is, hypotheses are generated using coarse and global information while details and local context information are utilised for verification and refinement. Initial hypotheses of parking vehicle rows are formed by detecting and filtering bright and dark ribbons in coarse scale. To this end, the differential geometric line extraction approach of Steger (1998) is employed, which is initialised with parameters derived from training data for the expected vehicle width and contrast. Then, single vehicles are reconstructed from these hypotheses by analysing grey value profiles perpendicular to each line. Individual vehicles are extracted by least-squares fitting of Gaussian kernels to the profiles. The decision about the existence of a vehicle is finally based on the resulting parameter values after fitting. Figure 8.11 illustrates different steps: ribbon extraction (left), kernel fitting (centre) and reconstruction of the selected vehicles (right).

In contrast to static regions, the distances between vehicles in slowly moving regions are not always close enough that the vehicle pattern forms a line in coarser scales. A better alternative is therefore to exploit the time-variance of the grey values in the regions with movements in a similar way as for free-flowing traffic. To this end, the spatio-temporal (3D) gradients are calculated

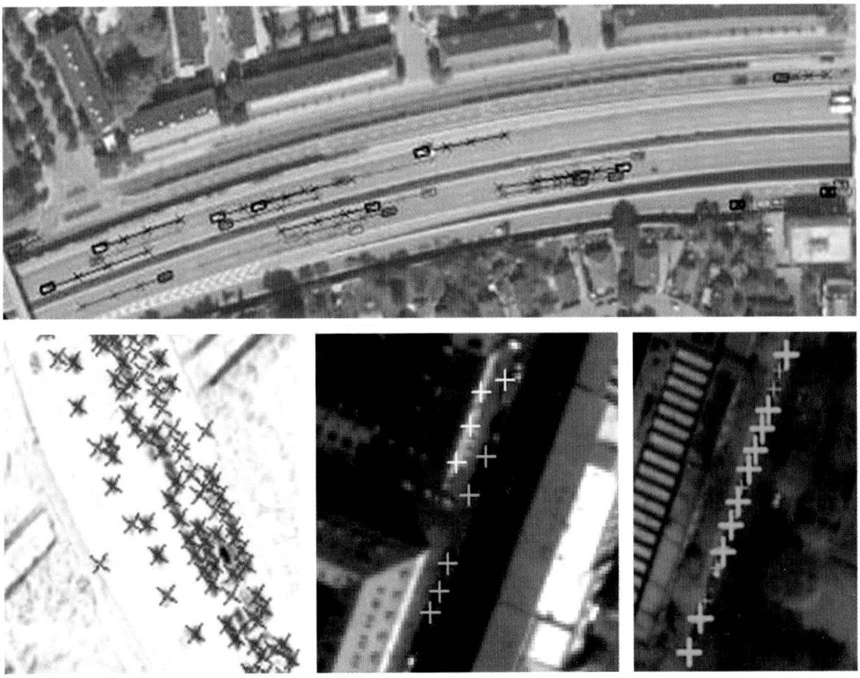

Figure 8.12 Top: Result of airborne monitoring of free-flowing traffic. Bottom left: Detected vehicles based on temporal grey value variance in congestion (cf. Figure 8.5). Bottom centre and right: Detected parked cars in satellite images (correct detections, cyan; missed hits, green; false alarms, white).1.12 Top: Result of airborne monitoring of free flowing traffic. Bottom left: Detected vehicles based on temporal grey value variance in congestion (cf. Figure 8.5). Bottom centre and right: Detected parking cars in satellite images; correct detections (cyan), missed hits (green), false alarms (white).

and segmented, resulting in the finally extracted moving vehicles. An example result of this algorithm is given in Figure 8.12, bottom left.

8.4 Evaluation and Discussion

At the end of processing, individual vehicles have been extracted for free-flowing traffic, traffic congestions and traffic jams. Figure 8.12 illustrates further examples for free-flowing traffic and traffic congestion monitored with airborne cameras, as well as parked cars detected in satellite imagery. Although some of the modules leave much room for technical and methodical improvements, numerous tests with different input data have been carried out already. To this end, ground truth data have been acquired by digitising cars and car tracks in satellite and aerial imagery. The following qualitative and quantitative conclusions can be drawn from the evaluation.

8.4.1 Evaluation of Results Using Satellite Images

Despite a ground sampling distance of approximately 0.6 m in the pan channel, vehicle detection in high-resolution satellite images remains a very challenging task. If setting the system parameters in such a way that the false alarm rate is sufficiently low (e.g. below 20%), the detection rate for single vehicles extracted from monocular satellite images lies in the range of 35–50%. A certain gain is to be expected when single-pass stereo information is available, because temporal features are also exploited. Since only one QuickBird single-pass stereo image is available to us, a statistical evaluation of the improvement does not make sense up to now. Yet the detection rates become much better for vehicle lines by reaching a completeness of 50–60%. This emphasises the importance of incorporating global knowledge about groupings of vehicles, as mentioned above.

The estimation of vehicle movements seems accurate and very reliable. For all correctly extracted vehicles, manual inspection has shown that the motion vectors have been derived correctly; that is, the cars' motion directions are approximately aligned with the road orientation and the computed velocities fall in the range of 45–60 km/h, which is plausible for city streets. It has to be mentioned that area-wide velocity estimation is not possible, as only a small fraction of 4% of the detected vehicles show the necessary colour fringes (Leitloff, 2011).

8.4.2 Evaluation of Results Based on Aerial Image Sequences

Vehicle detection rates in aerial image sequences are typically higher than for satellite images. Although the gain in spatial resolution regarding the panchromatic information amounts only to 20–30 cm for the medium-format cameras used in the test flights, the additional high-resolution multi-spectral information plays the most important role. Detection rates of at least 60% and in some cases up to 90% can be reached on highways and freeways. Empirical tests have shown that a higher false alarm rate than for satellite images can be allowed (e.g. 50%), because the multiple-hypotheses tracking is able to reject most of the false hypotheses due to the vehicle motion model employed. The number of cars lost during tracking – because of occlusions, for instance – is close to zero, largely because of the big advantage of using overhead imagery.

8.4.3 Estimation of Traffic Flow Parameters

Another issue worth investigation is the estimation accuracy of traffic parameters derived from vehicle extraction and velocity estimation. In particular, velocity profiles per lane are of special interest, as they allow reasoning about important traffic parameters like average velocity and traffic flow. To estimate the accuracy of velocity profiles derived from extraction results as given above, Monte-Carlo simulations with varying detection and false alarm rates were conducted. These were based on real ground truth data, from which a certain

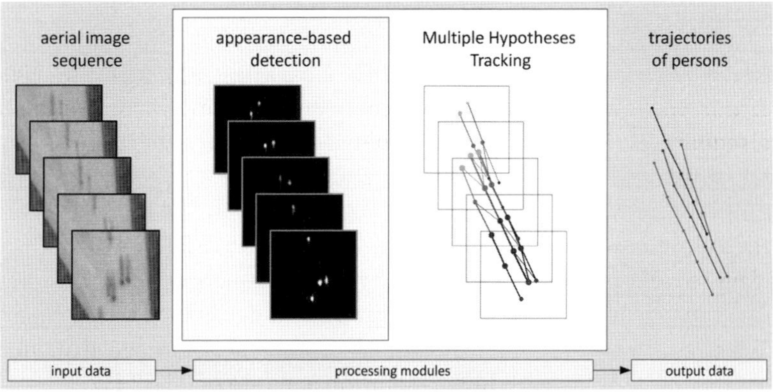

Figure 9.2 Overview of the processing strategy to detect and track persons in aerial image sequences.

artifacts induced by the process of image rectification and alignment further complicates the estimation of people's movement.

In the presented application we are interested in the trajectories of all persons in the observation area. They enable a subsequent microscopic analysis of the person's behaviour and their interactions. We have therefore set up the following processing strategy to detect and track individual persons by means of aerial image sequences (Figure 9.2). Every frame of the sequence is first scanned for potential object positions by an appearance-based detection approach. The observations are then described probabilistically and fed into an MHT approach. Eventually we extract the person's trajectories from the most likely global hypothesis. The following two chapters explain the entire strategy in more detail.

9.3 Detection of Persons

As stated before, nearly every tracking approach relies on a detection algorithm that reduces the high complexity of the observation data to a limited number of measurements. This set of detections should include all objects and at the same time as few false alarms as possible. In the presented application we are facing difficult problems that make it hard to fulfil these requirements. The small object size and low signal-to-noise ratio cause the main challenges.

Single persons appear in aerial images as tiny circular dots with a size of about 10 pixels (Figure 9.3). They have almost no additional features and thus look very similar. We assume that the persons are only observed directly from above, which is true for most optical airborne sensors. This assumption limits the variability of the object appearance and eases the detection process because the side view does not have to be considered. The person's appearance is augmented by a shadow when the sun is shining. This can increase the recognisability of a

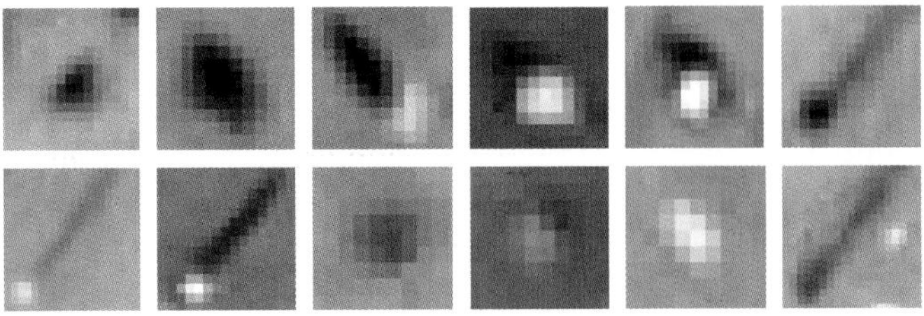

Figure 9.3 Examples for the visual variety of single persons in aerial images. Often the shadow is more visible than the person.

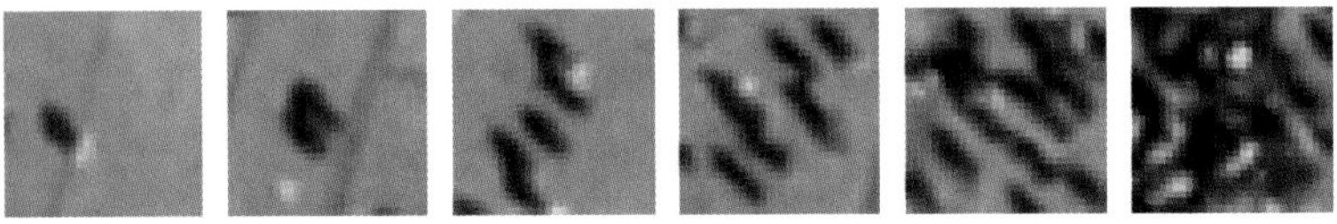

Figure 9.4 Effects of shadow and local object density on the recognisability of single persons.

single person significantly, but can also lead to a contrary effect in regions with higher object density (Figure 9.4).

9.3.1 Appearance-based Detection

As a consequence of the weak spectral signature of persons in aerial images, we have implemented an appearance-based detection approach based on local image features and a trained classifier. This detection method is much better suited under the given circumstances than any bottom-up data-driven approach such as moving object detection. An outline of the entire detection strategy is shown in Figure 9.5.

Every image of the sequence is processed independently. At first, some normalisation steps are conducted to eliminate systematic global variations. The image is scaled to a pixel size of 15 cm and turned so the persons' shadows always points in the same direction. These simple preprocessing steps reduce the object's visual variability, which again improves detection performance.

The implicit object model plays the most important part in the appearance-based detection approach. It is represented by the discriminative AdaBoost classifier (Freund and Schapire, 1997), which is trained prior to detection. For this task we manually collected about 2500 image samples of single persons, representing the total variety of the object class. Samples of the background class could afterwards be collected automatically in an iterative boot-strapping fashion (Schmidt and Hinz, 2011).

The classifier discriminates between persons and background by means of local image features. We allowed shape, colour and texture features to be extracted inside a detection window of 9 $\times$ 15 pixels. Its size is especially tailored to completely include a single person and parts of the potential shadow

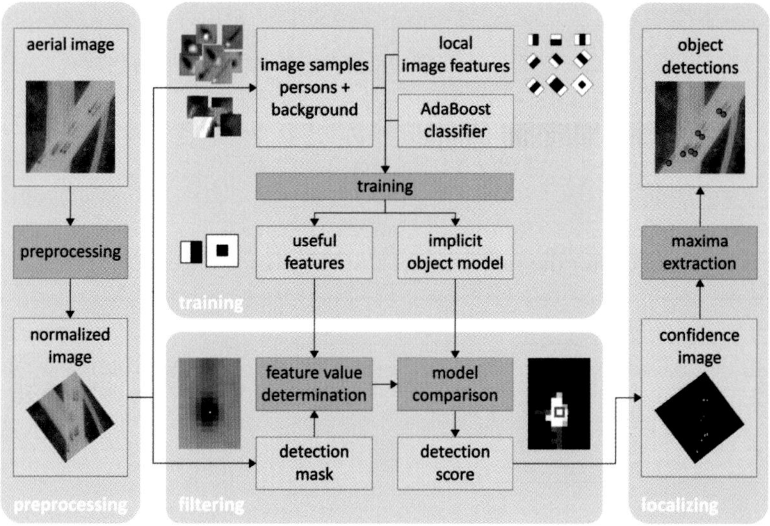

Figure 9.5 Overview of the appearance-based detection process with implicit object model.

(Figure 9.5). This approach is an efficient way to include the person's shadow into the implicit object model and take advantage of it during detection. During the design phase we only specified classes of image features and the size of the detector in which they are determined. This procedure led to a set of about 21,000 possible local image features. This number could be reduced tremendously to the 100 most useful features, because the performance of the AdaBoost classifiers already converges to its maximum when combining 100 thresholds.

Having the trained classifier and the set of useful image features completes the implicit object model. Searching objects in new images means evaluating every possible image position. We therefore conduct a brute-force search and run the detector over the entire aerial image. This results in a confidence image in which the pixel values are the detection score for the corresponding image position. Extracting local maxima in the confidence image finally reveals the potential location of persons. The corresponding confidence score can be used to classify the results into true detections and false alarms before they are passed on to the tracking algorithm.

9.3.2 Stochastic Modeling of Detection Results

In this section we show how to model the detection results stochastically, which provides numerous advantages for detection and tracking. The common way would be to conduct a binary decision at the end of the detection process and classify all observation into false alarms and potential objects based on the detection score. The disadvantage of this approach is that valuable information about the detection confidence is no longer available during tracking. In an application in which a low detection threshold is necessary in order to minimise the number of misses and enable a tracking-by-detection approach, this information would help to handle the high number of false alarms. Another advantage of stochastic modelling is that it allows for an optimal Bayesian decision. This way we can also include *a priori* knowledge about both classes and not rely solely on the detection score. Finally, probabilistic detection results can be integrated seamlessly into the MHT approach.

In object detection we discriminate the two disjoint events: H_1, the object is present; H_0, the object is not present. Their probabilities always add up to one: $P(H_1) + P(H_0) = 1$. Given a detection D with a score s we can conduct an optimal decision for one or both events by applying the maximum *a posteriori* (MAP) method:

$$\hat{H}_{MAP} = \text{argmax}_{H_i} P(H_i|s, D)\,,\ H_i \in \{H_0, H_1\} \tag{9.1}$$

Applying Bayes' rule and dropping the insignificant data term $p(s, D)$, the *a posteriori* probability of both events can be rewritten as follows:

$$P(H_i|s, D) = \frac{p(s, D|H_i) \cdot P(H_i)}{p(s, D)} \propto p(s|D, H_i) \cdot P(D|H_i) \cdot P(H_i) \tag{9.2}$$

Now it is composed of the *a priori* probability $P(H_i)$, the detection probability $P(D|H_i)$ and the detection score probability $p(s|D, H_i)$. All three terms have to be defined for both events in order to conduct a MAP decision. The *a priori* probability depends on context information and prior knowledge. It can be set to 1 in case none thereof is available. The detection probability depends on the quality of the detection method and the difficulty of the observed scene. Although the latter usually varies across the observation area, it is common to use a single value whose range can be estimated by evaluating different training sequences.

In contrast, the density distribution of the detection score probability of both events can be described analytically and estimated in advance. In order to do so we collected about 165,000 detection scores of false alarms and 2500 scores of correct detections. This was done by running the detection algorithm on images without persons and images containing object samples for the classifier training. The score distribution of both classes roughly follows a normal distribution, as can be seen in Figure 9.6. Because of the central limit theorem in probability theory, this observation should be similar for all application using an AdaBoost classifier with a high number of base classifiers.

The parameters of the distribution can be estimated with a least-squares regression. Because the detection score is a continuous value, it is advantageous to model the cumulative distribution first. The observation pairs (x_i, y_i) needed for the regression can be determined directly from the collected detection scores with $x_i = s_i$ and $y_i = p(S \leq s_i) = n(s \leq s_i)/N$, the ratio of samples

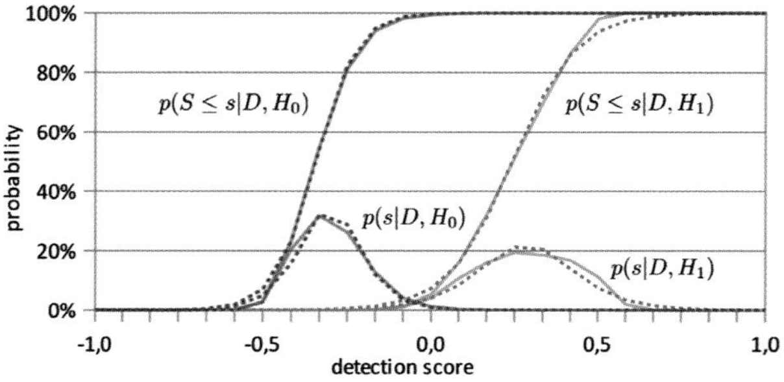

Figure 9.6 Distribution of the detection score for false alarms $p(s|D, H_0)$ and correct detections $p(s|D, H_1)$ with the described method for single person detection in aerial images. Presented are the true and cumulative distributions (solid line) as well as their sigmoid approximations (dashed line).

with a score lower or equal to the observation s_i. The cumulative distribution can be modelled by a sigmoid function:

$$p(S \leq s|D, H_i) \approx sig(s) = \frac{1}{2}[1 + \tanh(a \cdot s + b)] \tag{9.3}$$

Now the probability density distribution of the detection score follows directly by derivation of this equation and has the same parameters a and b:

$$p(s|D, H_i) \approx sig'(s) = \frac{a}{2}[1 - \tanh^2(a \cdot s + b)] \tag{9.4}$$

The sigmoid approximation follows the true distribution sufficiently well, as illustrated in Figure 9.6 and proven by a correlation coefficient of 0.999. It can therefore be used to calculate the *a posteriori* probabilities in Equation 9.2 and to conduct an optimal MAP decision.

9.4 Tracking of Persons

We use the MHT approach to generate trajectories from the image-wise detections. Facing the difficult challenges of our application, we decided to use one of the most sophisticated data association methods, which has been especially designed for such complex situations. In this section we first describe the used dynamic object model, then we present the probabilistic evaluation function and the general outline of the MHT approach. Finally we discuss some tracking results.

9.4.1 Dynamic Object Model

In Section 9.3 we presented the object model used to detect persons in aerial images. For tracking it is now necessary to define which parts are dynamic and which are static. For detection we use a single appearance model for all persons.

This is trained with many different object samples to capture the whole variety of persons' appearances in aerial images, hence making it generally deployable. For tracking it would be possible to derive an individual appearance model for every person thereof by iteratively training it with new detections. Yet the high similarity between different persons (Figure 9.1) and the time needed for training are reasons against this approach. We thus decided to work with a static appearance model that is equal for all persons.

The other part of the object model describes the position of a person. It is given not only in image coordinates but also in world coordinates, because the aerial images are transformed into geo-referenced ortho-photos. The position is of course a dynamic object feature. We model its variability with a near-constant motion model that assumes a smooth movement and neglects accelerations. A Bayesian framework for state estimation with a Kalman filter is used to determine the object's position and velocity from associated detections and to perform predictions to future points in time. The necessary values to characterise measurement noise and process noise are determined by analysing training sequences.

9.4.2 MHT Evaluation Function

The MHT approach tries to find the most likely hypothesis for the origin of all detections and for the dynamics of all objects over the entire observation period. In this section we present the probabilistic method that is used to evaluate different alternative hypotheses.

First, it is important to notice that a hypothesis in the MHT context has a different meaning than in other tracking approaches. Here, a single hypotheses Θ_i^t is always a global consistent explanation for all objects and all detection from the beginning of the observation until time t. It consists of a set of *association events* θ_k^t, one for each point in time: $\Theta_i^t = \{\theta_k^{1:t}\}$. The indices here and in the following text symbolise that a certain element has been selected from the available set at the given point in time. Each association event is a list of *assignment options*, which explains the origin of each detection and each object at the respective time t. The following four assignment options are possible:

- A detection is a false alarm (FA).
- A detection is from a new object (NT).
- A detection is from a previously detected object (DT).
- A tracked object is currently not detected (LT).

Having stated the basic terms and definition, the aim of MHT is to determine the hypothesis Θ_{MAP}^t with the the highest *a posteriori* probability of all possible hypotheses Θ^t given all measurements $Z^{1:t}$ for the entire observation period:

$$\Theta_{MAP}^t = \text{argmax}_{\Theta_i^t} P(\Theta_i^t | Z^{1:t}) \tag{9.5}$$

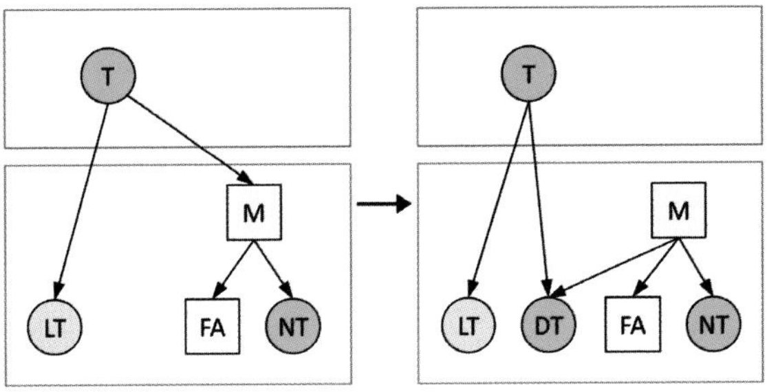

Figure 9.9 Left: Tracking graph after establishing all possible assignment options. Right: How the graph has to be altered if the DT option has been selected in one association event.

9.4.3.2 *Generating Hypotheses*

Before we can generate new hypotheses we have to complete the current association events by adding further assignment options to the tracking graph (Figure 9.9). All active objects are connected to an LT option and all new detections to NT and FA options. The established arcs are again weighted with the corresponding score. Now we form new hypotheses independently for every cluster using the method of Cox and Hingorani (1996) and the optimisations of Miller et al. (1997). This processing step is the most crucial one in the MHT approach. Although the sequential formulation of the hypothesis probability in Reid (1979) was the precondition to make MHT feasible for real-time tracking applications, practical implementations have suffered from the exponentially growing number of hypotheses. This is caused by the fact that every prior hypothesis generates as many new hypotheses as there are solutions to its current assignment problem (cf. Equation 9.7). The most effective method to limit the number of hypotheses to a manageable quantity was presented by Cox and Hingorani (1996). They transferred a well-known procedure to find a given number of best solutions to any assignment problem to the MHT approach. Now it became possible to precisely control and even adapt the number of tracked hypotheses as well as the complexity of the original MHT with a single parameter.

Having found the n most likely new hypotheses for every cluster, we succeed with post-processing steps. At first we process all assignment options in the current association events of the new hypotheses. For every DT option we correct the predicted object states with the assigned observation and alter the tracking graph according to Figure 9.9. Objects with an LT option are tracked further or become inactive depending on their status and track score (c.f. Section 9.4.2.3). For every NT option we initiate a new Kalman filter with the corresponding detection. Following this, the lists of active and inactive objects of every hypothesis can be updated. The tracking graph can also be cleansed of all unused knots and arcs by analysing its structure based on the current hypotheses. This procedure assures that the tracking graph always represents only the current hypotheses and their dependencies. Afterwards, existing clusters are split or set inactive if required.

At the end of each iteration we re-estimate the probabilities of occurrence for new objects and false alarms based on the current hypotheses (cf. Section 9.4.2). We also generate object tracks from the top hypothesis of every inactive cluster.

9.4.3.3 Clustering

Having presented the basic processing steps of our MHT implementation, we now go into more detail regarding the very important clustering procedure. Although the method to limit the number of hypotheses to a fixed quantity was a breakthrough for the efficient implementation of the original MHT approach, its full potential is only reached when combined with a clustering procedure. If we are able to split the tracking problem into, for example, five independent subproblems and calculate only the 10 best hypotheses for each, we can implicitly track the million best hypotheses for the entire observation area. The chance that this set includes the global optimal solution is therefore much higher than without clustering.

A single cluster is defined as a set of interacting hypotheses that have no dependencies with hypotheses of other clusters. Hypotheses interact if they claim the same detections. The clustering procedure was very complicated in the past (Roy et al., 1997). Yet, by using the data structure explained above, it reduces to a simple search for connected components in the graph. If hypotheses of different clusters claim the same detection, they have to be merged into a so-called supercluster. Interacting clusters can be found by looking for connected components in the tracking graph. Two clusters with the same component number are merged by combining their hypotheses. This can be done for pairs of hypotheses by adding up their scores and integrating their lists of active and inactive objects. The number of new hypotheses in a supercluster is the product of the number of hypotheses in the merged cluster. For this reason it can grow very quickly and has to be limited to the desired quantity. After merging, new clusters can be formed from all detections that have not been claimed by existing hypotheses. A cluster has to be split if it contains independent parts with different component numbers. This can happen whenever hypotheses are deleted and knots are removed from the tracking graph. A cluster is split by dividing its hypotheses individually into the new clusters. This procedure produces many identical new hypotheses, yet only unique ones may be added. A cluster can be deactivated if it no longer contains hypotheses with an active object. Before final deletion its top hypothesis is used to generate object tracks.

9.5 Experiments

In this section we analyse how the previously presented methods affect the results of tracking and give details on the runtime performance of the MHT algorithm.

Figure 9.10 A typical frame of each of the evaluated test sequences.

9.5.1 Data Sets and Evaluation Metric

We use six image sequences of two completely different tracking scenarios in the experiments to demonstrate the general usability of the proposed methods. Their characteristics and typical frames are displayed in Figure 9.10 and Figure 9.1.

The PETS sequences represent a common outdoor video surveillance setting and come from a well-known benchmark data set[1]. The publicly available detections of TU Darmstadt (Walk et al., 2010; Andriyenko and Schindler, 2011) are used as input for tracking and the provided complete ground truth as reference.

The other data set in our experiment contains aerial image sequences (AIS) which differ from surveillance videos in multiple ways (Table 9.1). They are taken from a moving aeroplane at a height of about 1.5 km. Each aerial image has been rectified, geo-referenced and cropped to show the moving persons on flat ground from the nadir direction in a defined RoI. The pixels have a size of only 12–15 cm on the ground, making it very challenging to detect single persons. To fulfil this task we designed an appearance-based detection module,

Table 9.1 Characteristics of the evaluated test sequences and the quality of the corresponding detection results (D) that are used as input in the experiments.

Name	Hz	Length	No. of objects/frame	Completeness-D	Correctness-D
PETS-S2L1	7	113 s	6	89%	76%
PETS-S2L2	7	62 s	23	56%	73%
PETS-S2L3	7	34 s	18	40%	72%
AIS1	2	7 s	90	71%	72%
AIS2	2	9 s	160	67%	50%
AIS3	2.5	12 s	200	76%	39%

which delivers position and score values for each detection (Schmidt and Hinz, 2011). The distribution of the score was modelled and integrated into tracking as outlined in Section 9.3.1. This was not possible for the PETS sequences because the original detector is not available to us. All AIS sequences as well as the manually collected reference trajectories are available on our web page[2].

The level of difficulty differs considerably between the test sequences. The varying number and density of persons generate numerous challenges for the tracking algorithm (Figure 9.10). Additionally, the detection results, serving as the basic input for MHT, are partially characterised by low correctness and low completeness (Table 9.1).

During tracking, each person is represented by the centre position of its bounding box and a corresponding velocity term. For PETS the person's height is also included. The necessary tracking design parameters as well as the Kalman filter parameters were determined in advance from the test sequences (PETS) or from independent evaluation sequences (AIS).

The CLEAR MOT metrics (Stiefelhagen et al., 2007) are used in this chapter to evaluate the tracking results. The *multiple object tracking accuracy* (MOTA) is a cumulative measure that is calculated by subtracting the false alarm rate (FA), the miss rate (Miss) and the mismatch rate (Mism) from 100%, the best possible result. For AIS the tracking result has to be closer than 50 cm (3–4 pixels) from the reference to establish correct correspondence. For PETS the bounding boxes must overlap by at least 50%. Beforehand the width of results and reference is set to a fixed aspect ratio of 0.41, as suggested by Dollar et al. (2012).

9.5.2 Number of Tracked Hypotheses

The first experiment answers the main question regarding the MHT approach: How does tracking multiple hypotheses affect the quality of the tracking results? For this purpose we ran MHT five times on each test sequence, varying the number of tracked hypotheses per cluster from 1 to 100. The results are listed in Table 9.2.

Table 9.2 Tracking results pursuing different numbers of hypotheses (#Hypo) per cluster. Values are given in percent and with respect to one reference result in order to draw attention from the overall results to the effect of the examined parameter.

PETS-S2L1					AIS1				
#Hypo	**MOTA**	**FA**	**Miss**	**Mism**	**#Hypo**	**MOTA**	**FA**	**Miss**	**Mism**
1	−80.6	−6.7	+87.8	−0.3	1 Ref.	32.9	4.1	56.9	6.2
2 Ref.	82.5	6.8	10.3	0.3	2	+14.6	−0.7	−13.4	−0.6
5	+0.4	+0.3	−0.7	−0.0	5	+13.4	+1.8	−14.4	−1.1
10	−0.5	+0.5	−0.0	+0.1	10	+15.5	+1.2	−14.7	−2.1
100	+1.7	−1.0	−0.7	−0.0	100	+17.6	+0.5	−15.3	−2.9
PETS-S2L2					**AIS2**				
#Hypo	**MOTA**	**FA**	**Miss**	**Mism**	**#Hypo**	**MOTA**	**FA**	**Miss**	**Mism**
1	−25.5	−3.8	+29.6	−0.4	1 Ref.	15.0	21.8	58.7	4.5
2 Ref.	29.6	5.5	64.4	0.5	2	+2.7	+2.3	−4.5	−0.5
5	+7.0	+0.3	−7.4	+0.1	5	+3.0	+3.1	−6.0	−0.0
10	+8.4	+0.8	−9.3	+0.1	10	+5.3	+2.5	−6.5	−1.2
100	+4.2	+0.2	−4.3	−0.0	100	+5.0	+3.1	−8.1	+0.1
PETS-S2L3					**AIS3**				
#Hypo	**MOTA**	**FA**	**Miss**	**Mism**	**#Hypo**	**MOTA**	**FA**	**Miss**	**Mism**
1	−7.7	+13.8	−8.3	+2.2	1 Ref.	25.1	22.9	43.5	8.5
2 Ref.	27.2	7.6	64.4	0.8	2	+1.0	+2.6	−2.7	−0.9
5	+3.6	−0.8	−3.0	+0.2	5	+1.3	+3.1	−4.7	+0.3
10	+2.2	+0.6	−3.1	+0.3	10	+3.2	+2.6	−5.0	−0.6
100	+0.3	−1.8	+1.6	−0.1	100	+2.0	+4.5	−5.8	−0.7

Starting with AIS, MHT reduces the miss and mismatch rate compared to *single hypothesis tracking* (SHT) in all three sequences while at the same time increasing the false alarm rate. This illustrates the main characteristic of MHT, the ability to improve tracking, especially in difficult situations, by pursuing multiple hypotheses. However, this feature leads also to a higher false alarm rate because persistent false detections are tracked much better too. This situation is most obvious in AIS2 and AIS3, which contain a very high number of false detections (50–61%, Table 9.1). The chance that these false alarms are persistent over multiple frames is therefore much higher, leading to more false tracks.

In the first two PETS sequences almost all persons are missed when tracking only one hypothesis, as opposed to the third sequence, which has a high false alarm rate.The reason for these poor results is a domination of either the false alarm option (FA) or the new target option (NT). This situation can occur when there are more false alarms than new targets at the beginning of the sequence as in PETS1 and PETS2, or vice versa as in PETS3, and only a single hypothesis is used to determine the new target and false alarm density. A weak initial hypothesis can then reinforce itself during tracking. The SHT results of the AIS sequences are not corrupted thereby, although they contain a much higher number of false alarms. This is because here the score of each detection is integrated into tracking, compensating errors made during density estimation.

Comparing the different variants of MHT, this shows that more hypotheses lead to better results. However, this trend is not always as distinct as in AIS1 and, often, most of the improvements can already be achieved with a low number of hypotheses. Thus it depends on the tracking scenario how many hypotheses should be pursued until the results converge. Clustering allows this number to be kept very low. Tracking two hypotheses per cluster in PETS1 actually means 3200 global hypotheses per frame, five means 52 million, ten means 720 million and 100 means 1.8 billion.

The overall tracking results are relatively low except for the first PETS sequence. This reveals the high influence of detection on the final tracking results. The tracking algorithm reduces the number of false detections significantly in all sequences. However, if recall is low, it becomes very hard for tracks to collect enough evidence to pass the confirmation threshold.

9.5.3 Density Estimation

The proposed method for automatic density estimation of false alarms and new targets is analysed in the following experiment. We conducted multiple runs keeping the density uniform and the quantities fixed or estimating one or both automatically. The necessary numbers of new targets and false alarms have been estimated *a priori* from the reference and detection results. The evaluation was done tracking 10 hypotheses and using a temporal window of 2–3 s. The results are found in Table 9.3.

The highest MOT (Multiple Object Tracking) accuracy is achieved in almost all sequences using the proposed method. This shows that the estimated density distribution and the number of NT and FA better reflect the tracking scenario than the *a priori* determined fixed values. The amount of improvement depends on the sequence.

It stands out that estimating only the number of FA and NT leads to a higher false alarm rate and a lower miss rate. The reason for this is that some false detections are always tracked, leading to an underestimation of the number of FA and an overestimation of NT. This effect can be seen most clearly in AIS2 and AIS3, where lots of persistent false detections are present.

In general, automatic estimation of the density distribution leads to fewer false alarms and mismatches, and more misses. In PETS2 and PETS3 the increase in the miss rate is exceptionally large. Both sequences are characterized by high object density and many missed detections. This causes many

Table 9.3 Tracking results using fixed (f) uniform or automatically (a) estimated probability density (D) and fixed or automatic number (N) of new targets and false alarms. Results are given in % with first row as reference (cf. Table 9.5.2).

PETS-S2L1						**AIS1**					
D	**N**	**MOTA**	**FA**	**Miss**	**Mism**	**D**	**N**	**MOTA**	**FA**	**Miss**	**Mism**
f	f	78.1	12.6	8.9	0.4	f	f	45.0	6.8	38.1	10.1
a	f	+5.6	−6.4	+0.9	−0.1	a	f	+3.5	−0.5	+3.4	−6.4
f	a	−2.3	+2.1	+0.2	+0.0	f	a	+2.0	−0.3	+0.3	−2.0
a	a	+3.9	−5.3	+1.4	+0.0	a	a	+3.4	−1.5	+4.1	−6.0
PETS-S2L2						**AIS2**					
D	**N**	**MOTA**	**FA**	**Miss**	**Mism**	**D**	**N**	**MOTA**	**FA**	**Miss**	**Mism**
f	f	47.8	10.8	40.1	1.3	f	f	15.9	29.8	49.4	4.9
a	f	−6.1	−3.6	+10.3	−0.5	a	f	+3.4	−5.0	+2.6	−1.1
f	a	+0.2	+0.8	−1.0	+0.0	f	a	−1.1	+2.5	−2.1	+0.7
a	a	−9.8	−4.5	+15.0	−0.7	a	a	+4.4	−5.5	+2.8	−1.6
PETS-S2L3						**AIS3**					
D	**N**	**MOTA**	**FA**	**Miss**	**Mism**	**D**	**N**	**MOTA**	**FA**	**Miss**	**Mism**
f	f	29.8	10.4	58.7	1.1	f	f	26.0	23.2	43.7	7.1
a	f	−6.1	−2.6	+9.1	−0.3	a	f	+2.8	−2.9	+1.3	−1.3
f	a	+0.4	−0.1	−0.6	+0.3	f	a	−2.1	+6.6	−7.3	+2.8
a	a	−0.4	−2.2	+2.6	−0.0	a	a	+2.2	+2.3	−5.2	+0.8

short tracks, which quickly fall below the deletion threshold and contribute to the false alarm density. Therefore the FA distribution increases, especially in regions where objects are located, making the track confirmation even more difficult. The results could be improved if the detection score were integrated into tracking of the PETS sequences. This would compensate for mistakes made during automatic density estimation, as can be seen in the AIS results.

Another aspect that has to be considered in this evaluation is the short length of the test sequences. Thus the ability of the estimation method to learn and improve is limited. For very short sequences a combination with *a priori* knowledge might lead to better results.

9.5.4 Detection Score Integration

The previous experiments have already indicated that integrating the detection score can lead to better and more stable tracking results. The following experiment has been carried out to analyse the benefits of this method in more detail using the AIS sequences. They have been processed with and without the

Table 9.4 Tracking with and without use of the detection score (DS). Results for different threshold (TH) settings.

			AIS1		AIS2		AIS3	
$pT(s > TH)$	TH	$pFA(s > TH)$	MOTA DS	MOTA	MOTA DS	MOTA	MOTA DS	MOTA
0.999	0.28	0.87	41.0	–	2.5	−3.3	−2.2	–
0.995	0.36	0.29	48.9	18.5	9.1	−13.9	13.3	−81.9
0.99	0.39	0.11	49.9	33.0	15.5	−5.4	22.2	−33.9
0.98	0.43	0.04	48.4	51.5	20.3	4.5	28.2	−6.6
0.95	0.48	0.007	43.3	48.5	17.6	10.3	27.3	23.7
0.90	0.51	0.002	30.6	30.6	9.8	9.7	−0.8	−0.8
0.80	0.56	0.0005	26.8	28.4	17.3	18.4	6.8	6.8

detection score. The detection threshold has also been varied. The results are listed in Table 9.4.

The first three columns show the relationship between the detection threshold and the score distribution of objects and false alarms, as displayed in Figure 9.6. On comparing the MOT accuracy of all three sequences when processed with and without detection score several aspects become obvious.

First, integrating the detection score makes it possible to operate with a low threshold. The results stay high, although significantly more false detections are introduced in the tracking process. Additionally, the best results are similar (AIS1) or higher (AIS2, AIS3) compared to tracking without the detection score.

Another hypothesis stated in Section 9.5.3 can also be confirmed. Tracking with the detection score leads to good results over a wider range of threshold settings. The dependency between detection threshold and tracking results could therefore be reduced significantly.

9.5.5 Runtime Performance

It is also an important aim of this chapter to bring more clarity into the discussion about MHT runtime performance. Although we know that the absolute runtime is affected very much by the hardware and software used, the following experiment nevertheless shows the order of magnitude one can expect when using the MHT approach.

We conduct the experiments on a standard PC using a single 2.67 GHz processor. Implementation has been done completely in C++. We assume given detections and measure only the time for tracking. Figure 9.11 displays the absolute runtime per frame for all test sequences, tracking a varying number of hypotheses per cluster.

The runtime increases with the number of tracked hypotheses as expected. There is also a clear dependency between runtime and the number of objects

A Process-based Model Approach to Predict Future Land-use Changes and Link Biodiversity with Soil Erosion in Chile

CHAPTER

10

Andreas Ch. Braun, Callum Banfield

10.1 Introduction

Biodiversity is alarmingly threatened by human activities in the 21st century (Brooks et al., 2002; Butchart et al., 2010; Sala et al., 2000). Many of the processes that cause biodiversity to decline are relate to land-use changes and especially to deforestation (Dale et al., 1994; Jetz et al., 2007; Poschlod et al., 2005; Reidsma et al., 2006). Species extinction rates are considered to be several orders of magnitude above the natural level (Barnosky et al., 2011; Pimm and Raven, 2000). Numerous lines of reasoning argue that this biodiversity loss cannot be accepted (Maier, 2013). As well as ethical reasoning that man has no right to drive others species to extinction, there are utilitarian arguments: humans gain important benefits from intact biodiversity (Diaz and Duffy, 2006; Mace et al., 2012; Rey-Benayas et al., 2009; Schläpfer et al., 1999). The *diversity stability* hypothesis argues that biodiversity stabilises the resistance and resilience of ecosystems. Thus, man can foster ecosystem functions and services by protecting biodiversity (Hooper et al., 2005; Kinzig et al., 2001; Loreau et al., 2002; Naeem et al., 2009; Schulze and Mooney, 1993).

Vice versa, man can also impose threats on human society by reducing diversity. Among several other impacts, it has been shown that in many ecosystems, reducing biodiversity increases the threat of soil erosion. Merely punctual declines in biodiversity like, such as cutting down a few hectares of forest, may not have significant effect. However, by altering biodiversity at the landscape level, for example, by largely replacing diverse forests with less diverse land-use types, may induce severe soil degradation processes in entire landscapes (Bonilla and Vidal, 2011; Bonilla et al., 2010; Endlicher, 1988; Mathieu et al., 2007). Thus, by altering biodiversity at the landscape level, man may create a spatial configuration that not only affects environmental systems but also has threatening reciprocal effects on human society by increasing soil erosion.

Monitoring and modelling relevant land-use processes is crucial to assess potential impacts. Predictions of future land-use processes can be considered as visualisations of these spatial configurations that threaten coupled human–environmental systems. For these reasons, a large body of research has been dedicated to applying remote sensing to (1) land-use mapping (Green et al., 1994; Krisp et al., 2013; Ramankutty, 2010; Treitz and Rogan, 2004) and (2) biodiversity assessments (Gillespie et al., 2008; Innes and Koch, 1998; Kerr and Ostrovsky, 2003; Nagendra, 2001; Turner et al., 2003). However, few works have been published to link both and relate them to soil erosion. This contribution represents an approach to draw a link from land-use to biodiversity and erosion as impacts of reduced biodiversity. For this purpose, biodiversity is understood as species diversity of ecosystems and structural diversity, for example, the density of vegetation in the understorey strata. It shows how future land-use changes related to biodiversity can be modelled and how remote sensing can be applied to estimate erosion risks. Such estimates are then a starting point to assess erosion risks *in situ*. Hence, this approach combines fieldwork and remote sensing and outlines how fieldwork can be a starting point for remote sensing analyses, which in turn are then a focal point for further fieldwork.

10.2 Approach of this Study

This contribution exemplifies the relationship between land-use change processes, biodiversity and implications in a case study. With respect to remote sensing it explains how techniques can (1) be applied to map land-use change *processes* instead of merely mapping land-use *classes*, (2) use (remotely sensed) environmental data to predict future land-use changes and (3) link land-use change to ecological issues like biodiversity and ecosystem functions (erosion protection). It focuses on a study site in central Chile. Here, land-use change has been intense since the liberal turn in the 1970s. Land-use change has considerably affected biodiversity, and the decline of biodiversity has increased soil erosion[1].

The analysis provides three methodological steps. First, it outlines how transition processes can be mapped using advanced learning methods from remote sensing. Second, it is demonstrated how environmental data can be used to predict future land-use changes. Third, predicted land-use changes will be related to environmental impact and erosion processes.

The study uses rather old data sets (1975 and 1986), which is unusual for remote sensing studies. The rationale for doing so is that (1) the major change in the landscape has taken place between 1975 and 1986 and (2) that by using these data, the results of the predicted land-use changes can be validated using more recent data. The prediction of land-use change processes visualises one aspect of the spatial configuration that may impose a biodiversity decline in Chile.

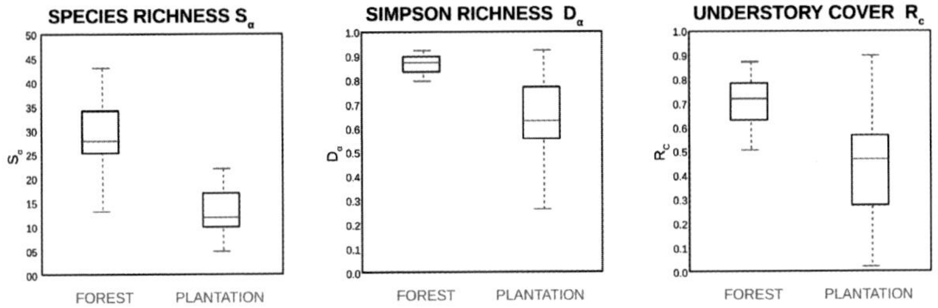

(a) Transition prob. (b) Prediction (c) Observation

Figure 10.1 Impact of land-use intensity on biodiversity. Land-use intensity increases from forests to shrublands and plantations. Biodiversity, measured in terms of species richness, true alpha-diversity and understorey cover, decreased with increasing intensity.

10.3 Process-based Land-use Assessment

In 2011, 2012 and 2013, 175 plots were sampled Braun (2013) to build a biodiversity database in central Chile. They represent Braun–Blanquet assessments and incorporate the number of plant species (species richness) and the coverage of vegetation in the plots. An analysis of this database shows that diversity is considerably reduced in some land-use types (plantations) in comparison to more natural land-use types (shrublands) and most of all, in comparison to native forests. For instance, the number of species is significantly reduced by up to 68% in native *Nothofagus obliqua* forests, than in commercial *Pinus radiata* plantations contain only 32% of the plant species typically ocurring in native Nothofagus obliqua forests (statistically significant according to t-test, $p < 0.001$). Fig. 10.1 shows that it is not only species richness, but also structural diversity and the coverage of vegetation *below* the canopy that are reduced. Hence, it has to be assumed that land-use intensity has a considerable negative impact on diversity (cf. Braun (2013)). On the basis of this finding, the first analytical step should consist of a multi-temporal description of land-use changes in the study site. However, the study site not only consists of three land-use types (i.e. plantations, forests and shrublands) but is complex mosaic of different land-use and land-cover classes. In Braun et al. (2014) 14 different land-use classes had to be considered in three small subsets. A straightforward approach would be to map each of the 14 classes in each of the images and compare maps afterwards (post-classification analysis). Considering 14 different land-use classes and taking into account that for each couple, land-use class A may change to class B and vice versa, $14 \times 14 = 196$ land-use change processes are theoretically possible. However, among those, many processes are irrelevant for biodiversity and the threats associated with declines in biodiversity (e.g. a change from residential to industrial should not alter erosion threats associated to biodiversity)[2]. Thus, mapping all relevant land-use types and then comparing maps afterwards will create unnecessary analytical complexity which is one of the major challenges in land-use threat governance Renn and Klinke (2013). In order to facilitate planning processes for erosion threat reduction, a more comprehensive mapping approach is to be developed, which focuses only on the relevant land-use change processes and ignores irrelevant change processes.

As stated in Part I, Chapter 5, a central issue is which entities are to be defined as *objects* during the mapping process. Post-classification analysis defines the *land-use types* as the objects of the analysis (e.g. forests or plantations are mapped as objects). A different approach is not to define land-use types, but *land-use change processes* as objects of the analysis (e.g. the *transition* from forests to plantations is mapped as an object). Considering change processes as objects is not straightforward, and few works have been dedicated to such approaches so far. Such an approach needs adequate mapping techniques. A framework for such analyses has been proposed by Camps-Valls et al. (2006, 2008). Their framework is based on change detection kernels K_{CD}^{t2-t1}. Each kernel is based on two images Im, in the case presented herein, Im^{75} and Im^{86}. Data from these images X^{75} and X^{86} are used as input data to compute kernel matrices (cf. Part I, Chapter 4). Various types of kernel have been proposed by Camps-Valls et al. (2006, 2008); here, the *image difference kernel* is used:

$$K_{ID}^{86-75}(x_i, x_j) = K_b^{86}(x_i^{86}, x_j^{86}) + K_b^{75}(x_i^{75}, x_j^{75}) \\ - K_b^{86-75}(x_i^{86}, x_j^{75}) - K_b^{75-86}(x_i^{75}, x_j^{86}) \qquad (10.1)$$

K_{CD}^{t2-t1} is based on four *base kernels*[3] K_b^t. The first and second represent the characteristics of the images in 1975 and 1986, and the third and fourth represent the cross information between the two points in time. The latter model the change of the landscape as represented by the images.

Because K_{CD}^{t2-t1} models the *change* between the two points in time it is no longer meaningful to use it to map land-use types (e.g. forests or plantations) as objects. Consider two patches of forest $F1$ and $F2$. $F1$ remained forest between 1975 and 1986 while $F1$ was cut down and changed to plantation. While $F1$ and $F2$ would be represented by very similar kernel values in the first two base kernels, they would yield totally different values in the second two base kernels and thus, in K_{CD}^{t2-t1}. The change detection kernel K_{CD}^{t2-t1} cannot be used to map the individual classes as object, but it is perfectly adequate to map the transition between object classes between the two points in time âŁ“ which fulfils the criteria for a process-based mapping scheme described above. An support vector machine (SVM) is used to classify the data represented by K_{CD}^{t2-t1} and thus, map the transition processes between 1975 and 1986.

At this point, a method to distinguish processes relevant to biodiversity and its associated threats has been developed. However, one important challenge remains. Since the process-based SVM works without a rejection class Braun (2009), the entire image is assigned to one of the relevant processes (i.e. although areas that did not change their land-use type or changed in a process that is not relevant, are mapped). Thus, a method has to be developed that recognises which areas were not affected by biodiversity-relevant change processes. Again, defining all the other change processes as objects, mapping each, and declaring them as irrelevant afterwards significantly raises the complexity of the analysis. the the complexity, it is aimed to distinguish areas with irrelevant changes without explicitly defining them as objects. A mapping approach to accomplish this task is given by one-class SVM (Braun, 2010; Chen et al., 2001; Tax and Duin, 1999). One-class SVM does not discriminate between objects of type A and type B (i.e. relevant and irrelevant change processes herein) but discriminates objects of type A from the remainder of the data space (i.e. relevant processes from irrelevant processes). In order to do so, all relevant processes are unified within a single class (i.e. the relevant process class). Any other change (i.e. irrelevant) processes represented in K_{CD}^{t2-t1} have to be distinguished from this class. This is achieved by folding the data of the relevant process class into a sphere of arbitrary volume and arbitrary shape. Anything outside this sphere is considered irrelevant and excluded from the analysis. The final result thus emphasises only relevant change processes and outlines, where transitions have taken place that threaten biodiversity. With respect to biodiversity, any change between forests, shrublands and plantations is relevant, as biodiversity declines in the order already mentioned (cf. Braun (2013)). Hence, six processes are relevant. For each process, some training sites have been defined, where the respective process (e.g. a transition from shrubland to plantation) has occurred between 1975 to 1986. Then, these processes are represented by K_{CD}^{t2-t1} and mapped, using an SVM. As mentioned above, the entire study site is attributed to one of these processes (cf. Figure 10.2). Therefore, in a next step, irrelevant processes are excluded from the analysis using one-class SVM (cf. Figure 10.2).

As can be seen, in contrast to the complex mosaic of different land-use *classes*, the pattern of relevant land-use *changes* is relatively simple. The prevalent processes are replacement of forests and shrublands into plantations. Transitions from plantations into other classes rarely occur which is also true for change processes between forests and shrublands. The assumption of a predominant replacement process that transforms natural forests and shrublands into plantations is in good agreement with the findings of other authors Aguayo et al. (2009); Echeverria et al. (2006); Rojas et al. (2013).

10.4 Prediction of Future Land-use Changes

So far, the first analysis of the approach described in Section 10.2 has been accomplished. It is described how central Chile changed between 1975 and 1986. This description is purely phenomenological in the sense that no model knowledge (e.g. from theoretical land-use change models) has been used. Now, this phenomenological knowledge will be related to landscape patterns in order to analyse the properties of the areas that have changed. For this purpose, a process-based model will be developed that analyses the dependency of the

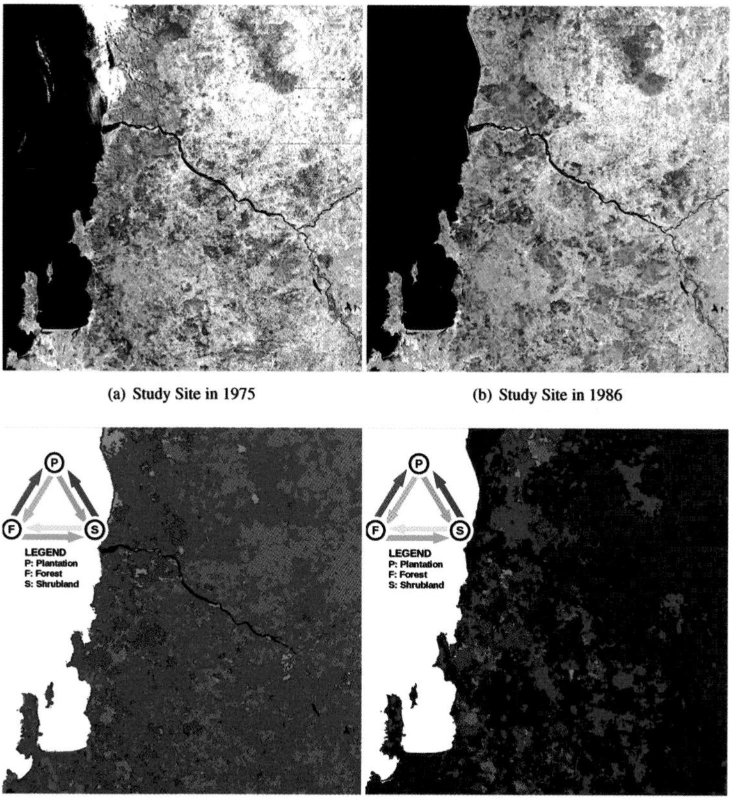

(a) Study Site in 1975 (b) Study Site in 1986

(c) Identified Change Processes (d) Relevant Change Processes

Figure 10.2 Mapped change processes before excluding irrelevant processes from the analysis and final result emphasising only relevant processes.

observed changes on environmental variables. A set of environmental variables has been collected in order to achieve this goal.

10.4.1 Environmental Parameters Used

A series of environmental parameters is used to predict deforestation processes in favour of the forestry industry. The parameters cover aspects of topography, ecosystems, climatology and land-use. For an overview, see Table 10.1.

10.4.1.1 Topographical Parameters

The set of topographical variables consists of five parameters, of which two were given by the geographical location of the study site (latitude and longitude) and three were derived from the digital elevation model (DEM) produced by the ASTER sensor (Fujisada et al., 2005; Hirano et al., 2003). A dependency of the land-use change processes on these data is expected, as deforestation for example, takes place in the forested coastal mountain range, in not too

Table 10.1 Environmental parameters used as predictive variables for deforestation in favour of plantations.

Environ. Aspect	Parameter	Data source
Topography	Latitude	USGS
	Longitude	USGS
	Elevation	ASTER-DEM
	Aspect	ASTER-DEM, AtCOR3
	Slope	ASTER-DEM, AtCOR3
Ecosystem	NDVI	USGS
	EVI	Jiang et al. (2008)
	Brightness	Tasselled Cap
	Greenness	Tasselled Cap
	Skyview	AtCOR3
	Shadow	AtCOR3
Climatology	Sum of annual precipitation	WorldClim
	Mean annual temperature	WorldClim
	Mean temperature July	WorldClim
	Mean Temp. Jan	WorldClim
	Minimum temperature July	WorldClim
	Maximum temperature January	WorldClim
Land-use	Percentage roads 2.25 km^2	*derived*
	Percentage urban 2.25 km^2	*derived*
	Percentage of agricultural 0.1 km^2	*derived*

steep terrain, while processes associated with the shrublands occur towards the interior of the country.

10.4.1.2 Ecosystem Parameters

For typical ecosystem parameters, such as primary productivity, coverage with vegetation and leaf area, surrogate data have to be used. Both the normalised difference vegetation index (NDVI) and the enhanced vegetation index (EVI) have been shown to be reliable proxy data for productivity and vegetation coverage, even when studying higher trophical levels (Pettorelli et al., 2006; Paruelo et al., 1997; Schloss et al., 1999; Sims et al., 2008). Furthermore, the brightness and greenness produced by the Tasselled Cap transformation are used, as this especially emphasises vegetation and soil properties in ecosystems (Dymond et al., 2002; Healey et al., 2005). Two features that relate to illumination condition – skyview and shadow as derived from atmospheric correction (Richter, 1997) – are also calculated. These parameters are relevant, since forest degradation, for example, is typically associated with losses in primary productivity. Furthermore, the establishment of secondary succession stages, such as the shrublands, depends on the openness and illumination of the habitat. In contrast, plantations are more tolerant to shade conditions.

10.4.1.3 Climatology Parameters

The WorldClim data by Hijmans et al. (2005) are used to reflect climatological conditions. The following parameters were used: sum of annual precipitation, mean annual temperature, mean temperature in July, mean temperature in January, minimum temperature in July, maximum temperature in January. These data are inherently linked to the growth conditions of different types of vegetation and therefore assumed to also be related to land-use change processes.

10.4.1.4 Land-use Parameters

The proximity to intensely used forms of land-cover has been shown to be major driver for future land-changes. In particular, the proximity to road, urbanised areas and agriculture is indicative. In order to reflect such proximity effects, the percentage of land-cover belonging to urbanised areas within the neighbourhood of 2.25 km^2, of agricultural areas (2.25 km^2) and of roads (0.1 km^2) has been computed for each pixel.

10.4.2 Model Development

These parameters are collected for the entire study site. They are used as predictive variables in a high-dimensional generalised linear model to predict the probability of land-use changes *from* the forest class *into* the plantation class. This transformation process has been shown to produce the most severe impacts on biodiversity (of up to 68% losses in species richness, cf. Braun (2013)). The model is based on kernel representations of the data in high-dimensional spaces. Within the high-dimensional spaces, it is linear. However, in the input space (environmental variables) it is highly non-linear (cf. MacKay (1992), and Part I, Chapter 4 herein). Thus, it is apt to reflect on the complex relationships between environmental data and land-use changes.

The model is fitted and trained to predict transition probabilities of the labelled areas, and it predicted these probabilities with acceptable levels of accuracy ($RMSE \approx 0.2$). Then, the model is applied to the entire data set to provide an estimate for the probability of a transition to the plantation class in the near future. It is assumed that areas that had *not* changed until 1986 but have a transition probability of over 50%, are prone to change into the plantation class. Results are compared to the map actually observed by Braun (2013) for the year 2005. Results are visualised in Figure 10.3. As the results indicate, the model proves very useful to predict the land-use change from forests to plantations. Within the major part of the study site, the model has correctly predicted changes to plantations. There are, however, a few exceptions to this tendency. For instance, changes to the plantation class close on the peninsula in the south-west of the study site have not been predicted. Furthermore, the

then analyse actual erosion processes in the field. The analysis used herein has been used to identify a study site where, indeed, higher erosion rates were observed in plantations than in forests.

Each step of this analysis requires further developments. Land-use change observations can be improved by incorporating more complex data sets or algorithms (e.g. conditional random fields Braun et al. (2014)). Land-use change modelling and predictions can – and should – incorporate explicit human knowledge on land-use change drives (e.g. the market situation for forest products or governmental legislature on forestry); by doing so, phenomena like the non-representation of additional land-use drivers can be reduced. Erosion modelling using the rUSLE or similar models should be validated with field data to provide fully-quantitative estimates of annual soil losses. Furthermore, the link between diversity and annual soil loss should be elaborated and more explicitly incorporate further study sites.

With all that being true, this contribution nonetheless highlights provides some important aspects. It shows how remote sensing can be applied in a combined and cyclic framework, where field observations (diversity assessments) foster remote sensing analyses (land-use modelling) that are used to derive scenario analyses (erosion estimation) which in turn, provide further field work (erosion assessment). In such frameworks, remote sensing is used in an integrated and interdisciplinary fashion to provide information on relevant environmental phenomena, going some steps beyond the scope of mere method development itself.

References

Aguayo, M., Pauchard, A., Azocar, G., and Parra, O. (2009). Land use change in the south central Chile at the end of the 20th century: understanding the spatio-temporal dynamics of the landscape. *Revista Chilena de la Historia Natural*, 82(3):361–374.

Auerswald, K., Flacke, W., and Neufang, L. (1988). Räumlich differenzierende berechnung großmaßstäblicher erosionsprognosekarten – modellgrundlagen der dABAG. *Zeitschrift für Pflanzenernährung und Bodenkunde*, 151(1988):369–373.

Banfield, C. (2013). Bodenerosion in kommerziellen baumplantagen in südzentralchile – mit einem ausblick auf die pestizidbelastung von plantagen-oberböden. Master thesis, Karlsruhe Institute of Technology (KIT), Karlsruhe, Germany.

Barnosky, A., Matzke, N., Tomiya, S., Wogan, G., Swartz, B., Quental, T., Marshall, V., McGuire, J., Lindsey, E., and Maguire, K. (2011). Has the Earths sixth mass extinction already arrived? *Nature*, 471(7336):51–57.

Beasley, D. and Huggins, L. (1982). *ANSWERS – User's Manual*. Purdue University, West Lafayette, UK.

Bonilla, C. and Johnson, O. (2012). Soil erodibility mapping and its correlation with soil properties in Central Chile. *Geoderma*, 189–190(2012):116–123.

Bonilla, C., Reyes, J., and Magri, A. (2010). Water erosion prediction using the revised universal soil loss equation (rusle) in a gis framework, central chile. *Chilean Journal of Agricultural Research*, 70(1):159–169.

Bonilla, C. and Vidal, K. (2011). Rainfall erosivity in central chile. *Journal of Hydrology*, 410(1):126–133.

Braun, A. (2009). Untersuchungen zur klassifizierung von vegetation mittels support vector machines. Diploma thesis, Karlsruhe Institute of Technology (KIT), Karlsruhe, Germany.

Braun, A. (2010). Evaluation of one-class svm for pixel-based and segment-based classification in remote sensing. In *International Archives of the Photogrammetry, Remote Sensing and Spatial Information Sciences*. ISPRS.

Braun, A. (2013). *Eine fernerkundungsgestützte geoökologische Prozessanalyse zum Risikozusammenhang zwischen Landnutzung und Biodiversität an einem Beispiel aus Chile*. Ph.d. thesis, Karlsruhe Institute of Technology (KIT), Karlsruhe, Germany.

Braun, A., Weidner, U., Echeverria, C., Aguayo, M., Rojas, C., Niemeyer, J., Kosov, S., Rottensteiner, F., Bähr, H., and Hinz, S. (2014). Design of a spectral-spatial pattern recognition framework for risk assessments using landsat data – a case study in chile. *IEEE Journal of Selected Topics in Applied Earth Observations and Remote Sensing and Remote Sensing*, 7(3):917–928.

Brooks, T., Mittermeier, A., Mittermeier, C., DaFonseca, G., Rylands, A., Konstant, W., Flick, P., Pilgrim, J., Oldfield, S., and Magin, G. (2002). Habitat loss and extinction in the hotspots of biodiversity. *Conservation Biology*, 16(4):909–923.

Butchart, S., Walpole, M., Collen, B., van Strien, A., Scharlemann, J., Almond, R., Baillie, J., Bomhard, B., Brown, C., and Bruno, J. (2010). Global biodiversity: indicators of recent decline. *Science*, 328(5982):1164–1168.

Camps-Valls, G., Gomez-Chova, L., Muñoz-Mari, J., Alonso, L., Calpe-Maravilla, J., and Moreno, J. (2006). Multitemporal image classification and change detection with kernels. In *Image and Signal Processing for Remote Sensing XII* (Vol. 6365, p. 63650H). International Society for Optics and Photonics.

Camps-Valls, G., Gomez-Chova, L., Muñoz-Mari, J., Rojo-Alvarez, H., and Martinez-Ramon, M. (2008). Kernel-based framework for multitemporal and multisource remote sensing data classification and change detection. *IEEE Transactions on Geoscience and Remote Sensing*, 46(6):1822–1835.

Chen, Y., Zhou, X., and Huang, T. (2001). One-class svm for learning in image retrieval. In *IEEE International Conference on Image Processing*.

Dale, V., Pearson, S., Offerman, H., and O'Neill, R. (1994). Relating patterns of land-use change to faunal biodiversity in the central Amazon. *Conservation Biology*, 8(4):1027–1036.

Diaz, S. and Duffy, J. (2006). Biodiversity and ecosystem services. In: Emmett Duffy J (Ed) *Encyclopedia of Earth*. Environmental Information Coalition, National Council for Science and the Environment, Washington D.C. Electronic publication.

Dymond, C., Mladenoff, D., and Radeloff, V. (2002). Phenomenological differences in Tasseled Cap improve deciduous forest classification. *Remote Sensing of Environment*, 80(3):460–472.

Echeverria, C., Coomes, D., Salas, J., Rey-Benayas, J., Lara, A., and Newton, A. (2006). Rapid deforestation and forest fragmentation of Chilean temperate forests. *Biological Conservation*, 130(4):481–494.

Endlicher, W. (1988). *Geoökologische Untersuchungen zur Landschaftsdegradation im Küstenbergland von Concepcion (Chile).* Franz Steiner Verlag Wiesbaden, Stuttgart, Germany.

Ferro, V., Porto, P., and Yu, B. (1999). A comparative study of rainfall erosivity estimation for Southern Italy and southeasouth Australia. *Hydrology Science Journal,* 36(6):549–564.

Fujisada, H., Bailey, G., Kelly, G., Hara, S., and Abrams, M. (2005). ASTER DEM performance. *IEEE Transaction on Geoscience and Remote Sensing,* 43(12):2707–2714.

Gillespie, T., Foody, G., Rocchini, D., Giorgi, A., and Saatchi, S. (2008). Measuring and modelling biodiversity from space. *Progress in Physical Geography,* 32(2):203–221.

Green, K., Kempka, D., and Lackey, L. (1994). Using remote sensing to detect and monitor land-cover and land-use change. *Photogrammetric Engineering and Remote Sensing,* 60(3):331–337.

Healey, S., Cohen, W., Yang, Z., and Krankina, O. (2005). Comparison of Tasseled Cap-based Landsat data structures for use in forest disturbance detection. *Remote Sensing of Environment,* 97(3):301–310.

Hickey, R. (2000). Slope Angle and Slope Length Solutions for GIS. *Cartography,* 29(1):1–7.

Hijmans, R., Cameron, S., Parra, J., Jones, P., and Jarvis, A. (2005). Very high resolution interpolated climate ssurface for global land areas. *International Journal of Climatology,* 26:1965–1978.

Hirano, A., Welch, R., and Lang, H. (2003). Mapping from ASTER stereo image data: DEM validation and accuracy assessment. *ISPRS Journal of Photogrammetry and Remote Sensing,* 57(5):356–370.

Hooper, D., Chapin, F., and Ewel, J. (2005). Effects of biodiversity on ecosystem functioning: a consensus of current knowledge. *Ecological Monographs,* 75:3–35.

Innes, J. and Koch, B. (1998). Forest biodiversity and its assessment by remote sensing. *Global Ecology & Biogeography Letters,* 7(6):397–419.

Jetz, W., Wilcove, D., and Dobson, A. (2007). Projected impacts of climate and land-use change on the global diversity of birds. *PLoS Biology,* 5(6):157–165.

Jiang, Z., Huete, A., Didan, K., and Miura, T. (2008). Development of a two-band enhanced vegetation index without a blue band. *Remote Sensing of Environment,* 112(10):3833–3845.

Kerr, J. and Ostrovsky, M. (2003). From space to species: ecological applications for remote sensing. *Trends in Ecology and Evolution,* 18(6):299–305.

Kinzig, A., Pacala, S., and Tilman, D. (2001). *The functional consequences of biodiversity: empirical progress and theoretical extensions.* Princton University Press, Princeton, USA.

Kouli, M., Soupios, P., and Vallianatos, F. (2008). Soil erosion prediction using the Revised Universal Soil Loss Equation (RUSLE) in a GIS framework, Chania, NorthNorth, Crete, Greece. *Environmental Geology,* 8(9):130–145.

Krisp, J., Meng, L., and Stilla, U. (2013). *Earth Observation of Global Changes.* Springer, Heidelberg, Germany.

Laflen, J., Lane, L., and Foster, G. (1991). WEPP – a next generation of erosion prediction technology. *Journal of Soil and Water Conservation,* 46(1):34–38.

Loreau, M., Naeem, S., and Inchausti, P. (2002). *Biodiversity and ecosystem functioning: synthesis and perspectives.* Oxford University Press, Oxford, UK.

Lu, D., Li, G., Valladares, G., and Batisella, M. (2004). Mapping soil erosion in Rodonia, Brazilian Amazonia: Using RUSLE, remote sensing and GIS. *Land Degradation & Development*, 15(2004):499–512.

Ma, J., Xue, Y., Ma, C., and Wang, Z. (2003). A data fusion approach for soil erosion monitoring in the Upper Yangtze River Basin of China based on Universal Soil Loss Equation (USLE) model. *International Journal of Remote Sensing*, 24(23):4777–4789.

Mace, G., Norris, K., and Fitter, A. (2012). Biodiversity and ecosystem services: a multilayered relationship. *Trends in Ecology and Evolution*, 27(1):19–26.

MacKay, D. (1992). Bayesian Interpolation. *Neural Computation*, 4(3):415–447.

Maier, D. (2013). *What's so good about biodiversity?: A call for a better reasoning about natures value*. Springer, Heidelberg, Germany.

Mathieu, R., Cervelle, B., Remy, D., and Pouget, M. (2007). Field-based and spectral indcators for soil erosion mapping in semi-arid mediterranean environments (coastal cordillera of central chile). *Earth Surface Processes and Landforms*, 32(1):13–31.

Morgan, R., Quinton, J., Smith, R., Govers, G., Poesen, J., Auerswald, K., Chisci, G., Torri, D., and Styczen, M. (1998). The European Soil Erosion Model (EUROSEM): a dynamic approach for predicting sediment transport from fields and small catchments. *Earth Surface Processes and Landforms*, 23(6):527–544.

Naeem, S., Bunker, D., Hector, A., Loreau, M., and Perrings, C. (2009). *Biodiversity, Ecosystem Functioning & Human Wellbeing: an ecological and economic perspective*. Oxford University Press, Oxford, UK.

Nagendra, H. (2001). Using remote sensing to assess biodiversity. *International Journal of Remote Sensing*, 22(12):2377–2400.

Paruelo, J., Epstein, H., Lauenroth, W., and Burke, I. (1997). ANPP estimates from NDVI dor the central grassland region of the United States. *Ecology*, 78(3):953–958.

Pettorelli, N., Gaillaird, H., Mysterud, A., Duncan, P., Delorme, D., Van Laere, G., Toigo, C., and Klein, F. (2006). Using a proxy of plant productivity (NDVI) to find key periods of animal performance: the case of the roe deer. *Oikos*, 112(3):565–572.

Pimm, S. and Raven, P. (2000). Biodiversity: extinction by numbers. *Nature*, 403(6772): 843–845.

Poschlod, P., Bakker, J., and Kahmen, S. (2005). Changing land use and its impacts on biodiversity. *Basic and Applied Ecology*, 6(2):93–98.

Ramankutty, N. (2010). Global land-cover change: Recent progress, remaining challenges. In Lambin, E. and Geist, H., editors, *Land-Use and Land-Cover Change*. Springer, New York, USA.

Reidsma, P., Tekelenburg, T., Van der Berg, M., and Alkemade, R. (2006). Impacts of land-use change on biodiversity: an assessment of agricultural biodiversity in the european union. *Agriculture, Ecosytems & Environment*, 114(1):86–102.

Renard, K., Foster, G., Weesies, G., and Porter, J. (1991). RUSLE – revised universal soil loss equation. *Journal of Soil and Water Conservation*, 46(1):30–33.

Renn, O. and Klinke, A. (2013). Space matters! impacts for risk governance. In Müller-Mahn, D., editor, *The Spatial Dimension of Risk. How geography shapes the emergence of riskscapes*. Routledge, New York, USA.

Rey-Benayas, J., Newton, A., DIaz, A., and Bullock, J. (2009). Enhancement of biodiversity and ecosystem services by ecological restoration: a meta-analysis. *Science*, 325(5944):1121–1124.

modelling implicitly contains the assumption that the main contribution of scattering in such layover areas is induced by building walls and not by clutter in front of the building or by the overlayed part of the roof. This approach can be generalised to SAR tomography (Fornaro et al., 2003; Reigber and Moreira, 2000) if more than one interferometric pair of the same viewing direction is available. Zhu et al. (2009) show that deriving 3D information via tomographic analysis and statistical model selection can be adapted to pixelwise calculation of dense height maps of urban areas, thus linking the concepts of SAR tomography with persistent scatterer interferometry (Ferretti et al., 2001; Kampes, 2006).

11.4 Approach of Building Reconstruction

The TanDEM configuration of the satellites TerraSAR-X and TanDEM-X establishes a new opportunity to exploit the interferometric phase acquired during a single pass of the two satellites, which minimises the effect of temporal decorrelation in the interferometric phases. Since our analysis of building features (Section 11.2) concludes with a recommendation to exploit the interferometric phase signature in layover areas, we will draw benefit from these data. In the following, we first introduce the underlying building model, followed by a short overview of the whole reconstruction approach. Afterwards, all processing steps are discussed in detail.

11.4.1 Building Model

As demonstrated in previous work (e.g. Thiele et al. 2010a, 2010b), the appearance of characteristic building features depends on the building geometry and properties as well as on the direct neighbourhood. Hence, as preparatory work for the bottom-up process of building reconstruction, we analysed two test data sets for Paris and Berlin to define a proper building model. Three building types are easily detectable: industrial halls, tower buildings and apartment houses. Of course, in inner districts of cities the situation is often more complex and classification is difficult. Hence, we ignored such areas and focused on buildings with a right-angled footprint with a flat-roof, and whose signatures do not overlay each other. The assumed building model is shown in Figure 11.4, and is defined by the parameter vector $P = (l, b, h, \rho)$.

11.4.2 Workflow

The general workflow of our approach is depicted in Figure 11.5. Since the coverage of satellites is quite huge, we decided to subdivide the approach into two main parts (separated by a white dashed line). The upper part consists of

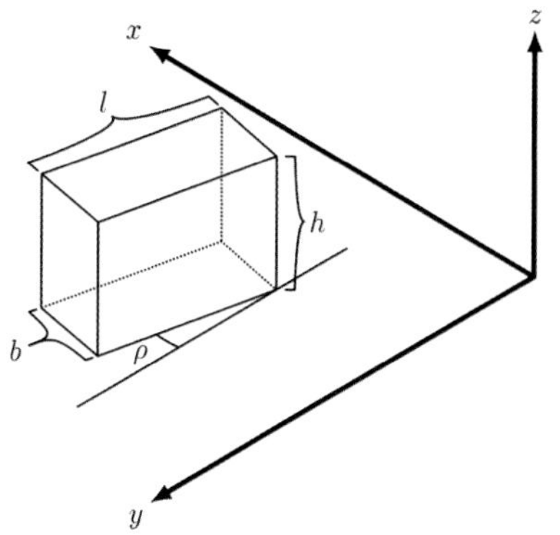

Figure 11.4 Building model.

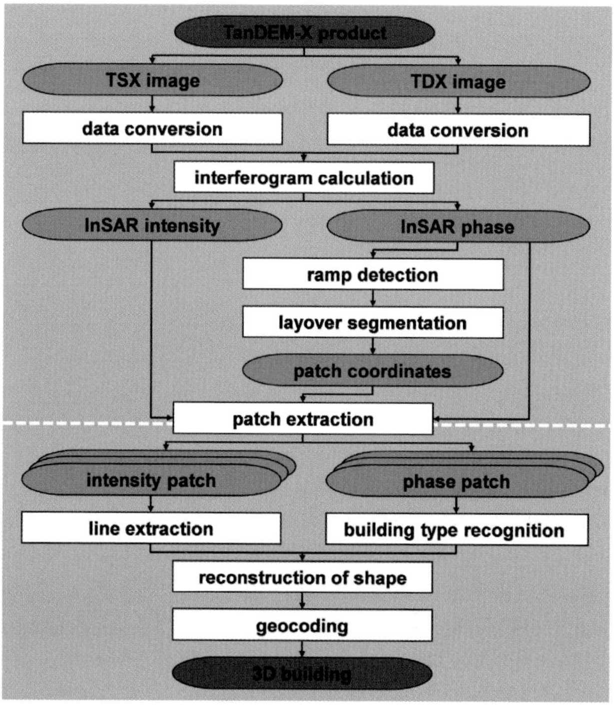

Figure 11.5 Workflow.

processing steps performed on the complete scene and the lower one of processing steps working on image patches containing a single building signature. This procedure allows more efficient data processing.

Focusing on the main steps of the workflow, pre-processing is first done, followed by applying the new phase ramp detector. Subsequently, image patches are segmented, containing a building hypothesis. Based on these patches, the

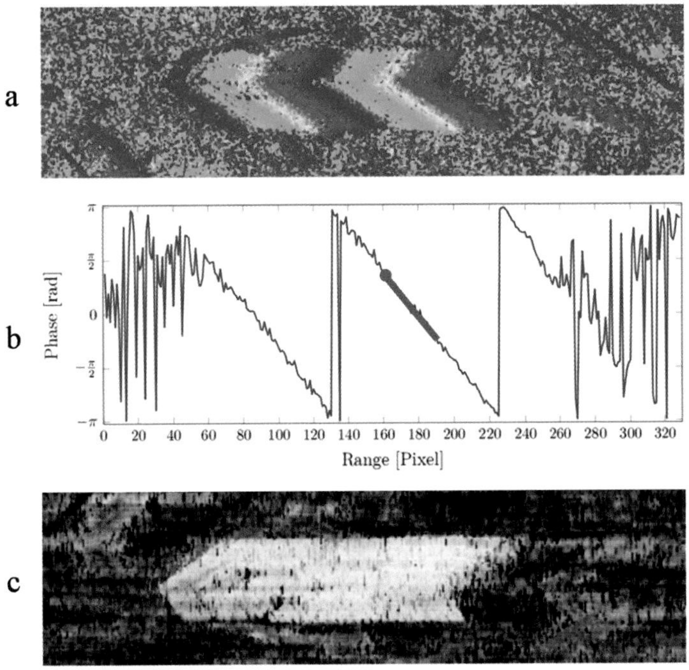

Figure 11.6 Detector of phase ramps: InSAR phase signature (a), phase profile overlaid with synthetic ramp (b, red), result of detector (c).

building type or the amount of visible building façades is determined. Finally, a detailed analysis is carried out to retrieve the parameters of the building, and its footprint is geocoded.

11.4.3 Pre-processing

The need for data pre-processing depends on the data. In our case, by using TanDEM data, the pre-processing starts with data conversion by importing CoSSC data files. More information concerning TanDEM-X data can be found in (Balss et al., 2012) and (Fritz et al., 2012). Afterwards, an interferogram calculation is necessary, because the originally provided interferograms are reduced in spatial resolution. A co-registration between master and slave image is not required because the data are provided co-registered. Finally, a flat-Earth correction is carried out to reduce phase fringes in the interferometric phases.

11.4.4 Detection of Phase Ramps

As already mentioned, building detection starts by analysing the full scene to expose layover areas that hint at building locations. An example of a fully developed layover area is given in Figure 11.6a. The corresponding phase profile is depicted in Figure 11.6b showing the typical saw tooth pattern, as the building height is about twice the unambiguous height Δh.

This observation of a remarkable phase pattern led us to generate a synthetic phase ramp as search pattern. Its gradient is defined by the parameters of the data pair – unambiguous height Δh, off-nadir look angle θ, and slant-range pixel spacing d_r. Since, for the given data, the variations of these parameters are negligible, the synthetic phase ramp is defined once and is thus similar for the whole scene. Finally, the implemented layover detector D determines the difference between interferometric phases (Figure 11.6b, blue line) and the synthetic phase ramp φ_s (Figure 11.6b, red line):

$$D[m,n] = \frac{-1}{L_{sr}} \cdot \sum_{k=1}^{L_{sr}} \left| W\left(\varphi_m[n+k] - \varphi_m[n]\right) - \varphi_s[k] \right|, \tag{11.8}$$

with

$$\varphi_s = k \cdot \frac{\Delta h \cdot \cos\theta}{d_r} \tag{11.9}$$

where m,n = position in the image
L_{sr} = length of synthetic phase ramp
φ_m = m-th row of interferogram
W = wrapped phase value in the interval $[-\pi, +\pi]$
k = pixel position in the synthetic phase ramp values $[1, L_{sr}]$

This comparison between real and synthetic phases is done twice – running from near to far range (see Figure 11.6b, red point at the left end of the red line) and vice versa to achieve sharper edges in the detector image. From these two intermediate results, the final detector result is obtained by applying a minimum operator. The detection result is visualised in Figure 11.6c, where bright areas characterise high correlation between real interferometric phases and the synthetic phase ramp. It can be seen that phase jumps in the layover signature do not affect the result. The tricky point using that detector is the choice of a proper length of the synthetic phase ramp.

11.4.5 Extraction of Building Patches

Two additional results of the phase ramp detector are shown in Figure 11.7a, which show high values (bright) in the layover areas. However, as the detector result is still noisy, the following post-processing is run. A simple thresholding of the result does not permit us to distinctly detect building areas. Thus, we filter the result with simple morphological opening and closing by reconstruction. This allows us to fill some areas within the building layover that were not detected as phase ramp. Furthermore, it smoothes the result without losing structure information. Results are shown in Figure 11.7b, where the image appears less noisy and the borders of the layover areas are sharpened.

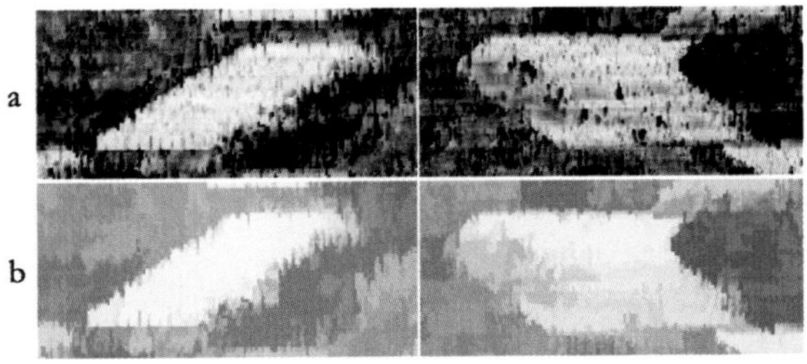

Figure 11.7 Result of (a) the phase ramp detector and (b) the morphological filtering.

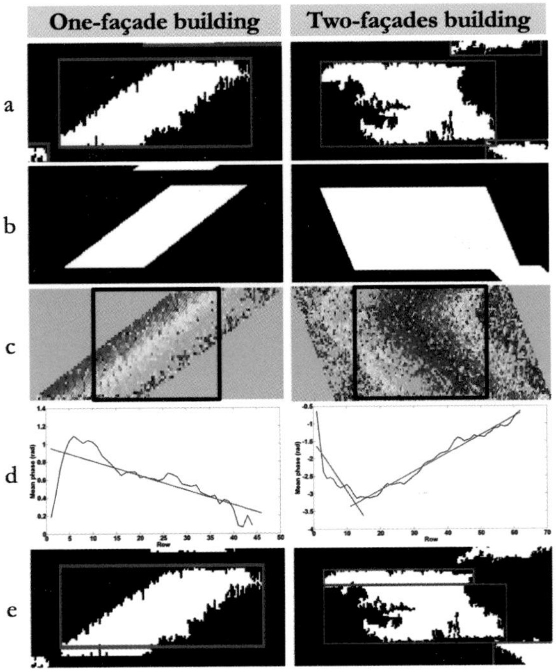

Figure 11.8 Steps to building extraction: (a) delimitation of building hypotheses; (b) parallelogram extraction; (c) parallelogram overlapped with phase image; (d) mean range phase values as a function of azimuth line; (e) new building delimitations.

Based on this morphologically filtered detector image, a threshold is applied in order to retain the building layover areas. For each connected component of the resulting binary image (see boxes in Figure 11.8a), a building hypothesis is created. For this purpose, a parallelogram shape is fitted by affine transform of a template the size of the original bounding box. Corrections are then brought to this transform by the least-squares method (LSM) until a threshold or a specific number of iterations is reached.

The results of this transformation are shown in Figure 11.8b. It can be seen that, for a layover area consisting of one façade (left column), and thus one parallelogram, the extracted boundaries fit well. Buildings with two apparent façades (right column) are well encompassed, yet with one parallelogram only. Thus, in the next step it is trivial to determine which type of building we have (one or two apparent façades), in order to separate the layover parts.

11.4.6 Extraction of Number of Building Façades

The number of visible façades is determined by analysing again the interferometric phase signature of the buildings. Figure 11.8c shows the phase values within each fitted parallelogram. Considering the one-façade building, the phase values decrease uniformly along the vertical direction. In contrast, for the two-façades building, phase values along the vertical direction first decrease, but increase again along the second façade. In order to bring this effect into light and thus detect the building edge, it is necessary to assess more than one vertical line due to the high phase noise. Thus, the phase values along the range direction within a specific area are summed (black boxes in Figure 11.8c). The area is chosen shorter than the completely detected parallelogram in order to avoid noise coming from building neighbourhood. Figure 11.8d shows for each row (horizontal axis) the mean of the summed phase values (vertical axis). For the two-façades building, a local minimum is observable, whereas the one-façade building shows a constant descending trend. Fitting lines (Figure 11.8d, red lines) through the obtained curves allow a final decision to be made between the two possible façade types.

For the separation of the two façades, the intersection point between the two fitted lines is used. That point indicates the position of the building edge. The two new bounding boxes are visualised in Figure 11.8e (red and blue parallelograms, right column). For the one-façade building, only one line is fitted and the original bounding box is adapted, considering the phase values at the box boundaries. Finally, a new parallelogram fitting by LSM is performed in each single bounding box, in order to extract single façade parameters.

11.4.7 Extraction of Building Parameters

For reconstruction of building shape, the four parameters of vector P are of high importance: length l, width w, height h and orientation ρ. They can all be directly derived from the extracted parallelograms. These are highlighted in Figure 11.9 for different building parallelograms.

The length l of the building can always be determined (Figure 11.9a), whereas the width w is only estimable if both façades contribute to the layover signature (Figure 11.9c). The building orientation is equal to the shearing parameter ρ of the parallelogram (Figure 11.9b). Building height h can be calculated twice: (1) using the parallelogram length and (2) using the underlying phase values. For this, the phase values are summed along the building orientation and the minimum and maximum values are searched. From their difference, the height can be deduced. At this point, a phase unwrapping is sometimes necessary.

11.4.8 Geocoding

Geocoding of the SAR data is accomplished within the Next ESA SAR Toolbox (NEST) by using Shuttle Radar Topography Mission (SRTM) height data. The

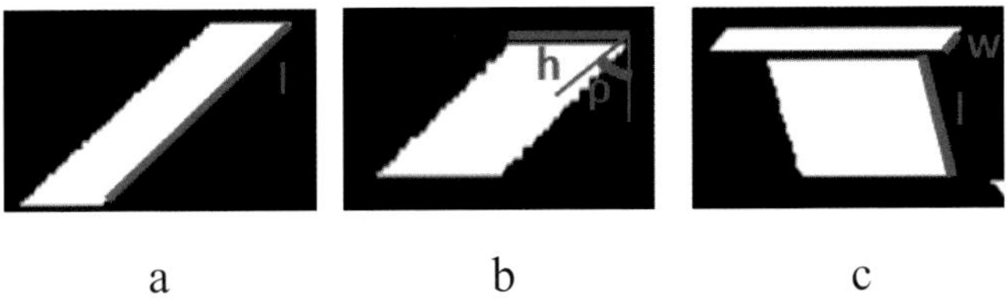

Figure 11.9 Extraction of building parameters: length l, width w, height h and orientation ρ.

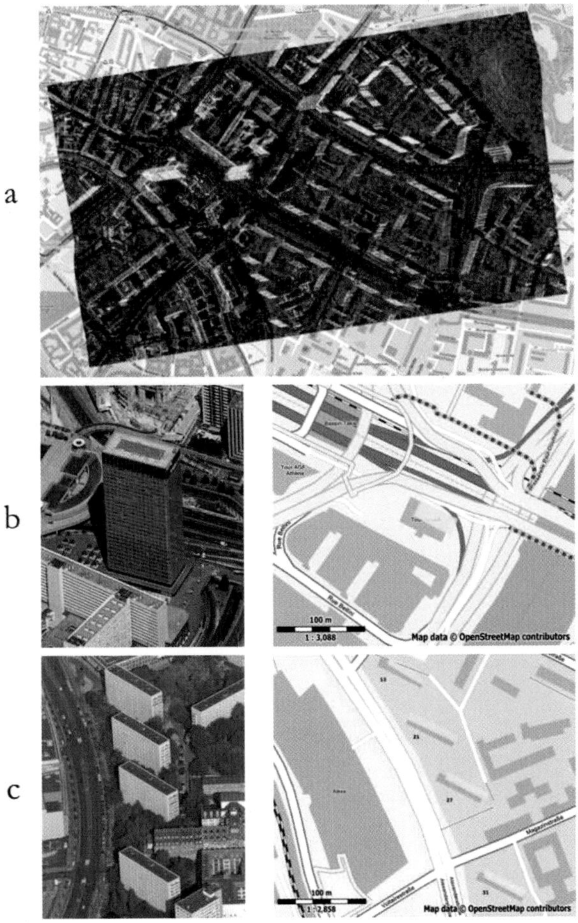

Figure 11.10 Geocoding map data overlaid with geocoded TerraSAR-X intensity image (a, Berlin), and optical data (source: Bing©) and map data overlaid with building footprints (Paris (b) and Berlin (c)).

result of the whole scene is given in Figure 11.10a. The four corner points of each building footprint are geocoded using the same height values. This has to be done to preserve the perpendicularity of the building footprint due to the indicated building model. In Figure 11.10b,c, geocoded building footprints are coloured yellow. Comparison with the map data shows an offset, probably caused by the quality of the SRTM data, which could be improved by considering actual local terrain heights.

11.5 Results

11.5.1 Data Sets

The high-resolution TanDEM-X science data are provided by the German Aerospace Center (DLR) as an additional product independent from the TanDEM mission to derive the global digital elevation model (DEM). The high-resolution SpotLight data shown in this chapter were taken over Paris and Berlin with a bandwidth of 300 MHz. They have comparable effective baselines of approximately 100 m, but, due to different off-nadir look angles, the unambignous heights are 52 m and 65 m. The pixel spacing is approximately 90 cm in the azimuth and 60 cm in the range direction. Furthermore, both data sets were recorded from ascending orbit at the beginning of January 2012. Hence, intersection effects due to strong vegetation are not given. In the following, detailed results are only shown for single buildings in the Berlin scene.

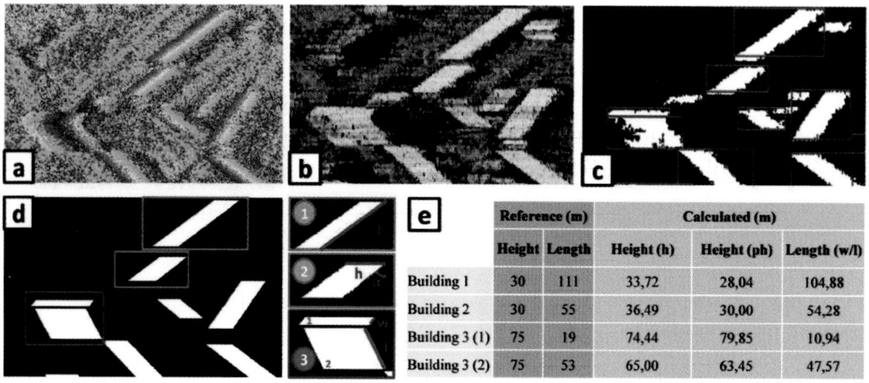

	Reference (m)		Calculated (m)		
	Height	Length	Height (h)	Height (ph)	Length (w/l)
Building 1	30	111	33,72	28,04	104,88
Building 2	30	55	36,49	30,00	54,28
Building 3 (1)	75	19	74,44	79,85	10,94
Building 3 (2)	75	53	65,00	63,45	47,57

Figure 11.11 Results of building extraction: (a) phase image, (b) phase ramp detector result, (c) building delimitations, (d) result of final parallelogram fitting, (e) extracted building parameters.

11.5.2 Evaluation of Building Parameters

Focusing on the results presented in Figure 11.11, it is recognisable that almost all buildings in the scene have been correctly detected and retained as building hypothesis. Considering parallelogram detection, it is obvious that our approach works well for both building types. The extracted building parameters are also satisfying. Additionally, the phase unwrapping step operates well, as for building 3. A satisfyingly accurate height has been extracted, despite the 2π phase ambiguity within the building signature.

In more detail, it is observable that the building lengths are slightly underestimated. There are two main possible reasons for this:

- The underestimation of the length of building 1 can be explained by the modification of the patch length before the last parallelogram fitting step, recognisable in Figure 11.11c.
- The width and length of building 3 are also underestimated, probably due to a slightly incorrect orientation of the fitted parallelogram's main axis.

Some similar remarks can be made concerning the height extraction:

- The height of the second part of building 3 is underestimated. This can be explained by the thresholding step after morphological filtering, whereby the resulting binary image is missing some pixels for correct reconstitution of the building shape (blue box Figure 11.11c).
- For some buildings, the difference in the extracted heights (based on parallelogram height or phase values) is about 5 m. This can be related to a slightly incorrect orientation of the fitted parallelogram and/or to a slightly incorrect position after fitting, which can already originate from the phase ramp detector.

More details, results and evaluations of this approach can be found in Dubois (2016) and Dubois et al. (2016).

11.6 Summary and Outlook

In this chapter, we have shown a new approach for building parameter estimation relying only on the analysis of building layover areas in the interferometric phase image of single-pass TanDEM-X data. In a first step, the new phase ramp detector is applied to the whole scene, followed by a segmentation step to obtain building patches. Based on these, the building parameters are extracted. Finally, the 3D building hypotheses are geocoded.

The phase ramp detector shows very good results that verify the high level of information given in the phase signature of buildings. Of course, only buildings showing a dominant layover can be extracted. Hence, to be able to also consider industrial halls or other buildings with less remarkable layover signatures, segmentation of roof areas in interferometric phases is necessary. A further advantage of this layover detection is the parameterisation of the ramp detector. Detailed studies on this are given in Dubois (2016) and Dubois et al. (2015, 2016), where recommendations to choose an appropriate length of the synthetic phase ramp and many further investigations on building extraction by applying InSAR and radargrammetric techniques can be found.

The phase values in layover areas are also influenced by the building roof type and building neighbourhood (Thiele et al. 2010b). In the future, the effect of this on the ramp detector should be analysed. Additionally, the already mentioned constant phase values along the building orientation should be considered in order to improve the layover detection and get rid of the remaining threshold-induced errors.

As a long-term objective, the combination of multi-aspect data can be helpful, as already shown for airborne configurations (Thiele et al., 2007).

References

Balss, U., Breit, H., Duque, S., Fritz, T., and Rossi, C. (2012). *TanDEM-X Payload Ground Segment: CoSSC Generation and Interferometric Considerations.* TD-PGS-TN-3129, Weßling: DLR Remote Sensing Technology Institute.

Barthelet, E., Mercier, G., Denise, L., and Reynaud, S. (2012). Feature based maximum likelihood model inversion for three-dimensional building extraction from single high resolution SAR images. In *Proceedings of IEEE International Geoscience and Remote Sensing Symposium, IGARSS 2012,* pages 3600–3603.

Bolter, R. and Leberl, F. (2000). Phenomenology-based and interferometry-guided building reconstruction from multiple SAR images. In *Proceedings of EUSAR 2000,* pages 687–690.

Brunner, D., Lemoine, G., and Bruzzone, L. (2010). Earthquake damage assessment of buildings using VHR optical and SAR imagery. *IEEE Transactions on Geoscience and Remote Sensing,* 48(5):2403–2420.

Dubois, C., Thiele, A., Welte, A., and Hinz, S. (2015). Parameterisation of layover detector for urban analysis in InSAR data. In *IEEE/ISPRS Joint Workshop on Remote Sensing and Data over Urban Areas.*

Dubois, C. (2016). Interferometric Synthetic Aperture RADAR and Radargrammetry towards the Categorization of Building Changes. PhD Thesis, Date: 11/11/2015, DOI(KIT): 10.5445/IR/1000051736.

Dubois, C., Thiele, A., and Hinz., S. (2016). Building Detection and Building Parameter Retrieval in InSAR Phase Images. *ISPRS Journal of Photogrammetry and Remote Sensing,* 114:228–241.

Ferretti, A., Prati, C., and Rocca, F. (2001). Permanent scatterers in SAR interferometry. *IEEE Transactions on Geoscience and Remote Sensing,* 39:8–20.

Ferro, A., Brunner, D., and Bruzzone, L. (2013). Automatic detection and reconstruction of building radar footprints from single VHR SAR images. *IEEE Transactions on Geoscience and Remote Sensing,* 51(2):935–952.

Fornaro, G., Serafino, F., and Soldovieri, F. (2003). Three-dimensional focusing with multipass SAR data. *IEEE Transactions on Geoscience and Remote Sensing*, 41:507–517.

Fritz, T., Duque, S., Bräutigam, B., and Bello, J. (2012). *TanDEM-X Ground Segment: TanDEM-X Experimental Product Description*. TD-GS-PS-3028, Weßling: DLR Remote Sensing Technology Institute.

Gamba, P. and Houshmand, B. (1999). Three dimensional urban characterization by IFSAR measurements. In *IEEE International Geoscience and Remote Sensing Symposium*, volume 5, pages 2401–2403.

Gamba, P. and Houshmand, B. (2000). Digital surface models and building extraction: A comparison of IFSAR and LIDAR data. *IEEE Transactions on Geoscience and Remote Sensing*, 38(4):1959–1968.

Gamba, P., Houshmand, B., and Saccani, M. (2000). Detection and extraction of buildings from interferometric SAR data. *IEEE Transactions on Geoscience and Remote Sensing*, 38(1):611–617.

Gernhardt, S. and Bamler, R. (2012). Deformation monitoring of single buildings using meter-resolution SAR data in PSI. *ISPRS Journal of Photogrammetry and Remote Sensing*, 73:68–79.

Kampes, B. (2006). *Radar Interferometry – The Persistent Scatterer Technique*. Springer-Verlag, Berlin, Germany.

Klausing, H. and Holpp, W. (1999). *Radar mit realer und synthetischer Apertur*. Oldenbourg Verlag, München, Germany.

Leberl, F. and Bolter, R. (2001). Building reconstruction from synthetic aperture radar images and interferometry. In Baltsavias, E., Grün, A., and van Gool, L., editors, *Automatic Extraction of Man-Made Objects from Aerial and Space Images (III)*. Balkema Publishers, Lisse, The Netherlands.

Quartulli, M. and Datcu, M. (2001). Bayesian model based city reconstruction from high resolution ISAR data. In *IEEE/ISPRS Joint Workshop on Remote Sensing and Data over Urban Areas*, pages 58–63.

Quartulli, M. and Datcu, M. (2003). Stochastic modelling for structure reconstruction from high-resolution SAR data. In *IEEE International Geoscience and Remote Sensing Symposium*, pages 4080–4082.

Reigber, A. and Moreira, A. (2000). First demonstration of airborne SAR tomography using multibaseline L-band data. *IEEE Transactions on Geoscience and Remote Sensing*, 38:2142–2152.

Sörgel, U., Thoennessen, U., Brenner, A., and Stilla, U. (2006). High resolution SAR data: new opportunities and challenges for the analysis of urban areas. In *IEE Proceedings Radar, Sonar and Navigation*, Vol. 153, pages 294–300.

Sörgel, U., Thoennessen, U., and Stilla, U. (2003). Iterative building reconstruction from multi-aspect InSAR data. *ISPRS Archives of Photogrammetry, Remote Sensing and Spatial Information Sciences*, 34.

Stilla, U. (2007). High resolution radar imaging of urban areas. In Fritsch, D., editor, *Photogrammetric Week*, pages 149–158. Wichmann-Verlag, Heidelberg, Germany.

Thiele, A., Hinz, S., and Cadario, E. (2010a). Combining GIS and InSAR data for 3D building reconstruction. In *Proceedings of 2010 IEEE International Geoscience and Remote Sensing Symposium, IGARSS 2010*, pages 2418–2421.

Thiele, A., Thoennessen, U., Cadario, E., Schulz, K., and Soergel, U. (2007). Building recognition from multi-aspect high resolution InSAR data in urban areas. *IEEE Transactions on Geoscience and Remote Sensing*, 45(11):3583–3593.

Thiele, A., Wegner, J., and Soergel, U. (2010b). Building reconstruction from multi-aspect InSAR data. In Soergel, U., editor, *Radar Remote Sensing of Urban Areas*, pages 187–214. Springer, Dordrecht, The Netherlands.

Tison, C., Tupin, F., and Maitre, H. (2007). A fusion scheme for joint retrieval of urban height map and classification from high resolution interferometric SAR images. *IEEE Transactions on Geoscience and Remote Sensing*, 45(2):495–505.

Tupin, F. (2003). Extraction of 3D information using overlay detection on SAR images. In *GRSS/ISPRS Joint Workshop on "Data Fusion and Remote Sensing over Urban Areas"*, pages 72–76.

Zhang, F., Shao, Y., Zhang, X., and Balz, T. (2011). Building L-shape footprint extraction from high resolution SAR image. In Stilla, U., Gamba, P., Juergens, C., and Maktav, D., editors, *JURSE 2011 – Joint Urban Remote Sensing Event*, pages 273–276. IEEE, New York, USA.

Zhang, Y., Sun, X., Thiele, A., and Hinz, S. (2015). Stochastic Geometrical Model and Monte Carlo Optimization Methods for Building Reconstruction from InSAR Data. *ISPRS Journal of Photogrammetry and Remote Sensing*, 108:49–61.

Zhu, X., Adam, N., Brcic, R., and Bamler, R. (2009). Space-borne high resolution SAR tomography: experiments in urban environment using TS-X data. In *Joint Urban Remote Sensing Event*, pages 1–8.

Detection and Classification of Collapsed Buildings after a Strong Earthquake by means of Laser Scanning and Image Analysis

CHAPTER **12**

Miriam Hommel, Thomas Vögtle

12.1 Motivation

A disaster – such as a strong earthquake – is an event that causes severe damage of various types and cannot be handled by the responsible organisation with its available resources. A worrying feature is the significant increase in the degree of damage observed in recent years (e.g. Rodriguez et al., 2009). The reasons for this are the ongoing growing in the human population, increasing building density and the increasing assignment of building areas in endangered regions. The problem after a strong earthquake is the inaccessibility of such areas. Streets are blocked by debris, surviving inhabitants react with panic, and fire or other phenomena may occur. It is thus difficult to obtain an overview of the general damage situation. Another aspect is the shortage of available resources for the rescue of buried people and the removal of debris. Finally, there is only a limited time period to rescue buried individuals alive, as the likelihood of finding entombed people alive decreases rapidly after 72 hours (Schweier, 2007).

The optimal planning and efficient allocation of rescue activities is thus essential. For this purpose information about collapsed buildings is necessary, as well as an estimation of the number of people probably trapped in such buildings and the resources needed for their rescue. With knowledge of these details, the available rescue forces and existing technical equipment can be distributed optimally to save as many people as possible.

To generate this information in the context of a disaster, fast data acquisition over large areas is necessary. In the case of the aftermath of a strong earthquake, only airborne approaches come into consideration. A flight shortly after an event using airborne laser scanning (ALS) seems to be most advantageous – compared to aerial images – due to the fact that it is an active sensor technology allowing data acquisition at night-time and under non-ideal weather conditions. This may be essential for a rapid response and appropriate decision-making. Also, smoke that is not too dense can be penetrated partly by a laser beam. This technology allows coverage of large areas in a short time (e.g. $>100\ km^2$ per day; Rathje and Adams, 2008), supported by the fact that 3D data are generated directly. In previous approaches, aerial or satellite images have often been used, but due to a lack of height information, several specific damage types could not be recognised.

The aim of this new approach is a fully automatic detection of collapsed buildings and a detailed classification of their damage type. From this information, a rough overview of the general damage situation should be derived as well as a detailed description of individual collapsed buildings as part of an efficient rescue measure. A method has also to be developed to model damaged buildings geometrically without common restrictions like orthogonality, parallelism, linear boundary structures or regular shapes, which may no longer exist after an earthquake. Subsequently, a specific detection and classification scheme of collapsed buildings must be created based on parameters such as changes in contour, surface, volume and height. To estimate the resulting quality a comprehensive verification has to be carried out. On the one hand this includes a comparison of damage types determined for collapsed buildings by the approach developed according to real damage types, and on the other hand it includes an investigation of unchanged buildings, which also have to be modelled and classified correctly. It also has to be analysed to what extent objects like vegetation or vehicles influence the results adversely. In a last step, the inclusion of complementary spectral data may lead to an improvement in damage classification.

12.2 Related Work

In this section we provide an overview about research activities concerning damage analysis based on remote sensing data (aerial, satellite and radar images) of disaster-affected areas. It focuses on approaches for detection and interpretation of collapsed buildings after earthquakes, while the effects of other natural disasters like floods and landslides are excluded. Some specific types of building damage, like crack formation or damage caused by fire, cannot be handled by airborne methods. The number of publications in the field of damage detection after earthquakes has significantly increased in recent years, confirming the importance and explosiveness of this subject. Existing methods can be distinguished in terms of the type of data, temporal resolution, classification principle and the number and type of damage classes that can be discriminated.

The types of data used for damage analyses comprise the type of the sensor as well as the products derived from the original data, and can include satellite images (e.g. Chesnel et al., 2007; Gusella et al., 2005; Huyck et al., 2004; Miura et al., 2007; Saito Spence, 2004), aerial images (e.g. Sumer and Turker, 2006), digital surface models (DSMs) derived from stereoscopic aerial images (Rezaeian and Gruen, 2007; Turker and Cetinkaya, 2005), DSMs derived from

- Surface structure (unchanged, erratic, with large planes)
- Size of the recognisable upper planes
- Volume reduction
- Total height difference compared to the initial height
- Recognisability of building contours
- Inclination change with reference to the initial situation within the building contour
- Damage of roof structure
- Debris structures outside building contour
  - Type of debris outside the building contour
  - Size of the debris outside the building contour
  - Additionally covered area outside the building contour
  - Slope of the piles of debris outside the building contour
  - Existing height at the building borders
- Existing damage elements

The definition of these quantitative and qualitative features is the result of analysing more than 1850 photographs of more than 150 collapsed buildings as well as the analysis of reports of rescue activities.

12.4 Applied Models

The problem of detection and recognition of objects in images or scanner data is the high complexity of the acquired scenes, which contain a variety of (other) objects with different occurrence. For this recognition, knowledge about the relevant objects and their relationships to one another is necessary, which is sampled, structured and represented in models. The model of an object may be devided into two parts, the *object model* and the *image model*. In the first, the characteristics of the object in the physical world are described, while in the latter the occurrence of the object in the data is represented. The transformation of the object model into the image space is reproduced by the *sensor model* containing the characteristics of the used sensor and the acquisition process. By means of this sensor model it is possible to predict the occurrence of the object model in the image space. Finally, knowledge about surrounding influences on the object as well as its relations to adjacent objects are defined in the *context model*.

Due to the fact that buildings after an earthquake may be damaged or still intact, it is necessary to create object and image models of collapsed but also of undestroyed buildings.

12.4.1 Models for Intact Buildings

An intact building consists of one or more building parts. Each of these is composed of a certain number of vertical walls and a roof that is an assembly of several roof areas and smaller objects like chimneys, dormers or antennas.

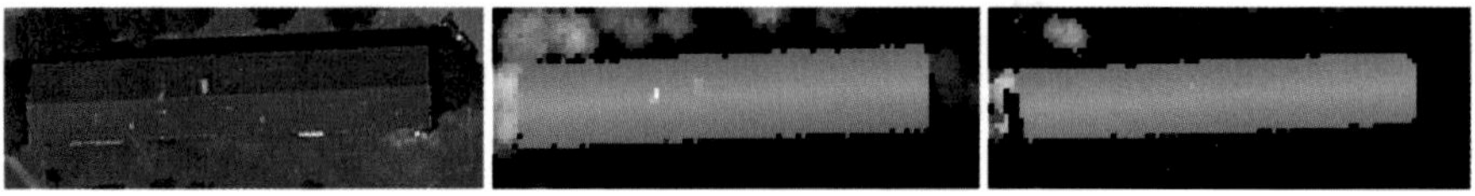

(a) (b) (c)

Figure 12.4 (a) Building in ortho-photo. (b) The same building in first echo DSM. (c) The same building in last echo DSM.

For the *object model* these roof areas are assumed to be planar and exceed a certain minimal size (e.g. 8 m^2). Therefore, curved surfaces like cupolas or barrel vaults do not meet the model assumption and must be approximated by smaller planar elements (but in practice the observed frequency of such roof shapes is very limited). The smaller objects on a roof can be treated as disturbing objects. The roof areas of buildings can be composed of different materials, e.g. roof tiles, concrete, metal, and even green roofs. Due to a common usage of buildings by humans, a minimal height of >2.5 m can be assumed.

The above-described components and features of an intact building are transformed to the data level (*image model*) by means of the *sensor model*. ALS data is used in raster format due to advantages such as easier and faster processing of regular data structure, noise reduction by robust determination of pixel height values, and others. Therefore, ALS data can also be treated like images. Both ALS data and (true) ortho-photos are based on a parallel projection and objects can be assumed (in general) not to be disturbed geometrically. Therefore, it is normally sufficient to scale the objects by an image scale (e.g. pixel size on the ground). Due to the 2.5D characteristics of ALS surface models, vertical walls are not explicitly represented in the image data. Hence, only features that can be derived from roof structures and building contours are used.

Planar roof areas occur in ALS surface models as homogeneous, planar regions with a certain minimum size (Figure 12.4 b,c). These areas generate only small first/last echo differences because of the smooth surface and moderate slope. Disturbing objects may be interpreted as outliers or separate regions that are smaller than the minimum size and can be excluded. In spectral data the colour and texture of roofs are caused by the materials used. Commonly, roof areas have a uniform colour and texture (Figure 12.4 a), while disturbing objects may lead to higher image textures.

12.4.2 Models for Collapsed Buildings

A collapsed building may feature different damage types as well as undamaged parts. Therefore, it is divided into intact parts and parts with specific damage types. While the components and features of intact parts are nearly the same as described in the last section for damaged parts, it is important to distinguish between features related to collapsed buildings in general and those related to changes compared to the original status. The geometric features of the different damage types have already been described in the classification scheme in Section 12.3. The general features of a collapsed building, for instance, contain roof parts broken into bigger (concrete) slabs, tilted walls with planar surface structure, or irregular surface structures caused by remaining wall parts or debris. Depending on the size of the rubble pieces, the damage type *debris* can be subdivided into three subclasses (Figure 12.2). Remaining parts of vertical walls (damage type 7c) can be characterised by larger height differences and a position inside the original building outline, whereas heaps of debris (damage types 7a, 7b) are often compact and show only small height differences. For all types of *debris* there will be a mixture of different materials in most cases. Features related to changes in building geometries include a reduction in volume, height differences compared to the original ones, changes in slope or debris outside the building contours. They have already been described in Section 12.3.

The damage type comprising broken bigger slabs and tilted walls occurs in an ALS surface model as relatively homogeneous, planar regions. In contrast, debris and remaining parts of vertical walls show an irregular height structure, with the latter additionally causing larger first/last echo differences as more or less linear structures. In spectral information, debris leads to high textures due to the mixture and granularity of the different materials.

12.4.3 Context Model

The *context model* in this application is focused on the relations of the relevant objects to adjacent objects in the scene (local context). This is important because buildings and other objects are often positioned at near distances, especially in urban areas. Therefore, recognition of collapsed buildings and their damage types may be difficult. For instance, vegetation (trees, bushes) located directly beside a collapsed building may be erroneously classified – based on a 3D surface model – as debris. This seeming increase in volume beside the building may lead to misclassifications. In consequence, not just the area of a building footprint has to be taken into account, but also the area around it. A special challenge is the correct assignment of objects or debris in the overlap of two areas to one of the related buildings.

12.5 Strategy for Damage Analysis

Due to the great variety of possible damage types and the fact that the underlying models are generic, a data-driven analysis strategy has been developed. After pre-processing, the data are divided into homogeneous areas (segments) with common features. For these segments, specific features are extracted from the data. By means of those features a classification is carried out that relates the segments or segment groups to the model. The whole strategy consists of three main parts. *Context analysis* is followed by the *generation of post-event building models,* which are derived solely from data acquired after the event

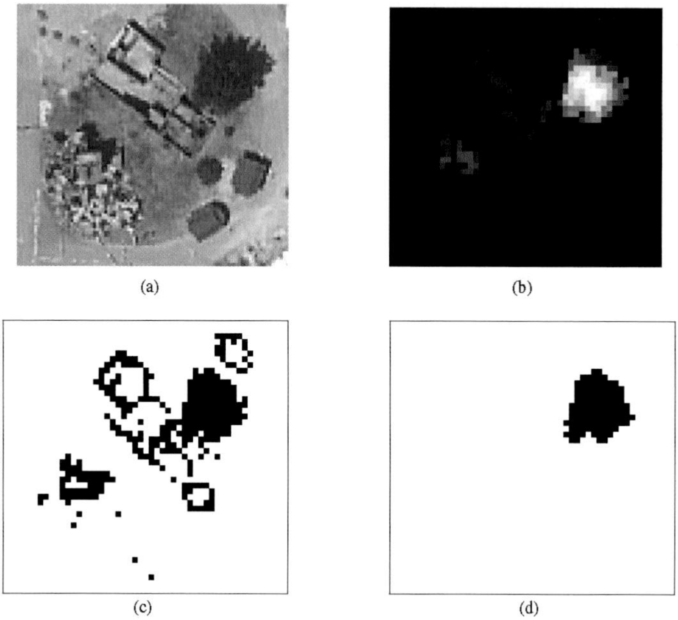

Figure 12.5 (a) Subset of an original RGB ortho-photo. (b) First/last echo differences. (c) Binarisation of first/last echo differences. (d) Elimination of line-shaped structures and elements inside the building area.

(mono-temporal). The last step is the *detection and detailed damage classification of collapsed buildings* based on a comparison of data before and after the earthquake.

12.5.1 Context Analysis

By means of the *context analysis* the local relationships between objects surrounding a building under investigation are determined. In this case, vegetation objects are extracted, which serve as exclusion areas for damage analysis (cf. Section 12.4.3). For the identification of vegetation objects two approaches have been developed. Their suitability depends on the degree of foliation and the type of available data.

For the first approach, ALS data (first/last echo) acquired after an event are used. To detect higher vegetation objects (trees, bushes, etc.) several parameters extracted from ALS data can be applied, including *surface roughness* (Brunn and Weidner, 1998), *height texture* or *first/last echo differences* (Tòvari and Voegtle, 2004; Rottensteiner et al., 2005), especially for differentiation between vegetation and buildings. While a laser beam can normally penetrate not too dense vegetation, which leads to larger first/last echo differences (Figure 12.5b), roof areas are solid and moderately tilted, resulting in small differences in the echoes. A problem occurs at the building borders where – due to the significant

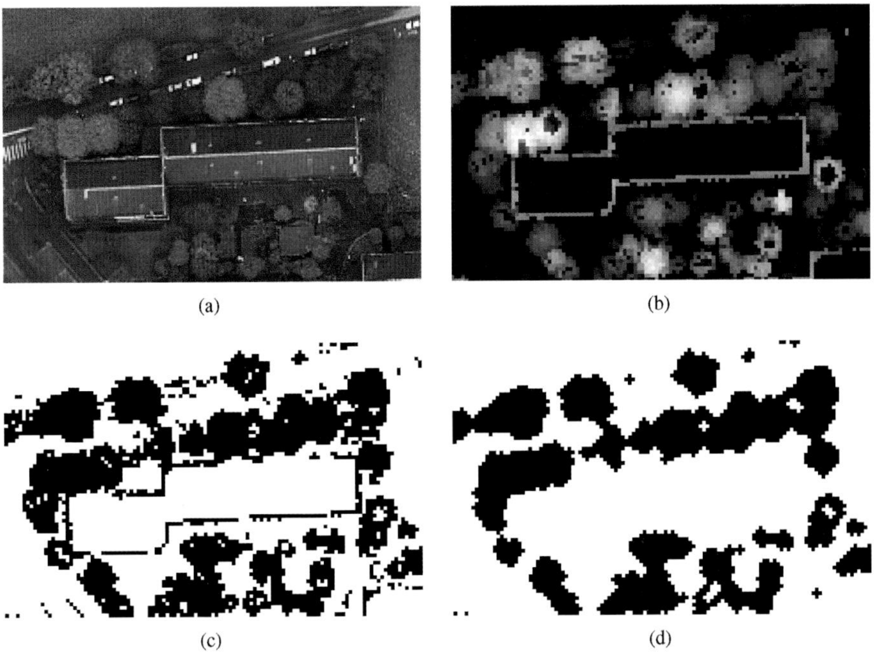

Figure 12.6 (a) Subset of an original CIR ortho-photo. (b) First/last echo differences where gaps can be observed inside dense vegetation. (c) Binarisation of first/last echo differences. (d) Resulting vegetation layer after applying mathematical morphology.

footprint size of the laser beam – a part of it may be reflected at the eaves and the other part at the ground, which also leads to large differences. In contrast to vegetation objects these differences are thin and line-shaped along the building contours (Figure 12.5c) and can therefore be eliminated easily, for example by mathematical morphology after a binarisation (Figure 12.5d). Rough debris and the remainder of walls may also lead to larger differences. These can be excluded due to their occurrence inside the building area (reference data) as it can be assumed that no vegetation could appear at such locations (Figure 12.5d). Debris outside the original building outline commonly has a structure that generates only small first/last echo differences.

Dense vegetation occurring in the middle part of a tree may also show small differences (Figure 12.6b), which may generate a gap in the area of the tree after binarisation (Figure 12.6c). Therefore, operations of mathematical morphology are applied to close these holes. After the operations described above, a final vegetation layer is created (Figure 12.6d) with the help of which vegetation areas can be eliminated from the data acquired after the earthquake.

If multi-spectral information is available (e.g. RGB or CIR aerial images) a second approach can be applied that is limited to daytime (illumination) and suitable weather conditions. An additional problem is the seasonal influence factor; that is, outside the time of foliation, trees are hard to recognise in aerial

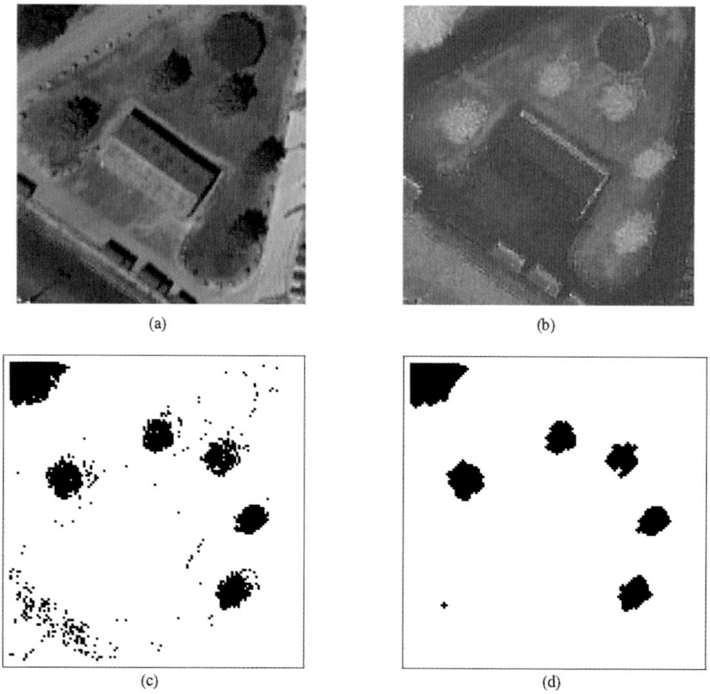

Figure 12.7 (a) Subset of an original CIR image. (b) Determined NDVI (bright = high NDVI). (c) Binarisation of NDVI and NIR values. (c) Elimination of isolated pixels and closing of small gaps by mathematical morphology.

images. The principle idea of this approach is the determination of a vegetation index. Due to the fact that leaves and foliage produce higher reflectance values in the spectral domain of visible green and near-infrared, the well-known Normalised Difference Vegetation Index (NDVI) has been established in the literature: $NDVI = (NIR - R)/(NIR + R)$. A problem is that in shadow areas, high NDVI values also occur. Therefore, to separate vegetation areas in multi-spectral images, two conditions have to be fulfilled:

$$NDVI > t_{NDVI} \quad \text{and} \quad NIR > t_{NIR}$$

where t_{NDVI} is a threshold for NDVI that has to be exceeded in order to be accepted as vegetation and t_{NIR} is the threshold for the minimum acceptable near-infrared value, NIR. Again, to eliminate small areas or isolated pixels and to close small gaps inside vegetation areas, a specific combination of operations of mathematical morphology is applied (Figure 12.7).

12.5.2 Generation of Geometric Building Models after a Disaster

The main task of this module is the generation of vector models (boundary representation) of buildings after an event. Before the extraction starts the

CHAPTER 13

A Settlement Process Analysis in Coastal Benin: Confronting Scarce Data Availability in Developing Countries

Andreas Ch. Braun, Ulrike Sturm-Hentschel, Stefan Hinz

13.1 Introduction

Coastal zones are the most densely populated and most dynamically changing areas in the world (Cicin-Sain and Knecht, 1998; McGranahan et al., 2007; Small and Nicholls, 2003), which is especially true for developing countries like the study area of this chapter – Benin. Migration processes from the hinterland and intrinsic population growth cause intense land surface pressure due to spontaneous and unplanned settlements and urbanization. Coastal areas, where anthropogenic change processes (e.g. settlement changes or deforestation) and natural processes (e.g. transgression and coastal erosion) interdependently co-occur, can be considered as coupled human-environment systems. The dynamics within such systems create degradation of ecological functions and assets on the environmental side (Burak et al., 2004; Gundlach and Hayes, 1978; Lee et al., 2006; Turner et al., 1996). On the human side, conflicts of interest arise (IPCC, 1996; Scheren et al., 2004). Urban sprawl and densification of existing settlements, in particular, are known to raise vulnerability of residents to many risks (Adger, 1999; Capobianco et al., 1999; Clark et al., 1998; Vogt et al., 2007).

In European, Australian or US coastal regions, anthropogenic change processes are usually regulated by governmental spatial planning, but the situation is much more complex and contingent in developing countries. Since many dwellings are constructed illegally, settlement boundaries advance out of the control of governmental spatial plans for risk reduction. Many residents of such dwellings may expose themselves to risks without being aware of it. For governmentally planned coastal management and the prevention of the aforementioned risks, it is crucial to gather information about ongoing processes, spatial demands and potential risks. However, gathering such information in the field is on the one hand too cost intensive to cover larger areas and on the other hand, too time consuming to keep pace with the rapid development.

At this point, remote sensing comes into play as it offers solutions to confront these constrictions. An experienced remote sensing operator can apply algorithms to map many times of the area that can be censused in the field within similar or even less time. However, settlement mapping based on remote sensing data in development countries is restricted by some serious constraints.

One of the most important constraints is scarce data availability. Unlike urban areas in Europe, Australia and the USA, which are covered by a wide array of remote sensors at least every couple of days, in developing countries, well-suited images are available only for a few points in time, often only every few years. This holds true even after the disclosure of the Landsat archives by USGS, since for settlement process studies finer scales are often relevant than can be assessed with the rather coarse spatial resolution of Landsat data. Besides very high spatial resolution data, multi-spectral and hyperspectral data are also rare, and Landsat is the only multispectral imager whose data is available back until the early 1970ies. Since settlement *process* studies inherently imply the comparison of several points in time, data availability issues aggravate the situation. In order to assess the densification of settlements or shifts of settlement boundaries, several points in time, represented by images of comparable spatial resolution, have to be analysed. These constraints often lead to situations in which data from more than a single sensor has to be used (Epstein et al., 2002; Jat et al., 2008).

Besides data availability, the particular characteristics of urbanised areas in many developing countries have to be taken into account. In contrast to analyses of cities in developed countries, settlement change cannot be assessed by mapping impervious surfaces. In many tropical and subtropical countries, impervious surfaces are rare, and often limited to roofs a and few roads. Even within densely settled urban areas, open soils and vegetation dominate the residential areas. This is the situation in the coastal area of Benin. Given these constraints, a mapping methodology has to be developed that is designed to achieve the following:

- *Detecting individual buildings* and not impervious surfaces for regional characteristics
- *Processing image data* with sufficiently high resolution to match the geographical scale of the process
- *Exploiting data* types acquired by various remote sensors to tackle data availability limits
- *Automatising mapping* of individual dwellings in order to be applicable to larger areas

Such a methodology is developed herein. Based on maps of individual dwellings produced by this methodology, changes in (1) the settlement boundary and (2) the settlement density are analysed.

13.2 Objectives of this Chapter

During recent years there has been increasing debate on the applicability of remote sensing to social and socio-economical processes (Dennis et al., 2005;

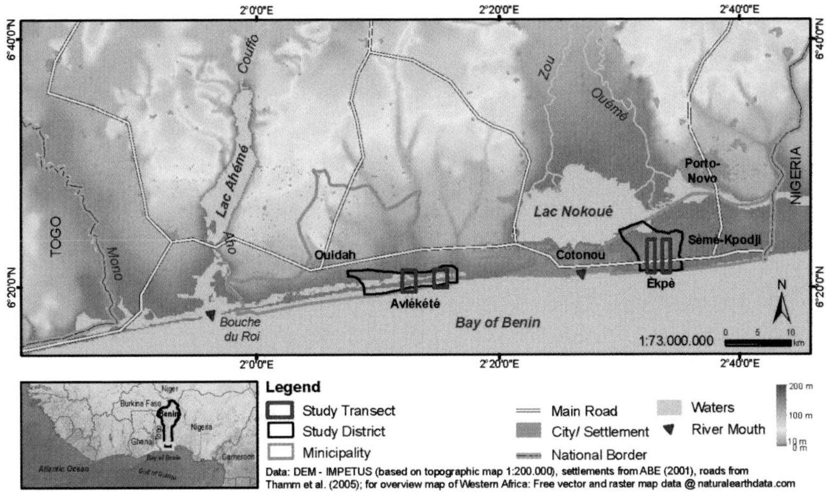

Figure 13.1 Benin, the coastal area and Èkpè

Lambin et al., 2009; Liverman et al., 1998; Liverman and Cuesta, 2008). One of the central issues within such approaches is the representation of space (in a geographical sense). While representing *relational spaces* as required for many studies (e.g. actor-based studies) is challenging with remote sensing, representing processes in container spaces is more promising. Fortunately, phenomenologically describing settlement change is possible in *container spaces*. Thus, remote sensing and photogrammetry have contributed a number of worthy approaches for this task. Nonetheless, automatically extracting man-made objects like roads and buildings and distinguishing them from optically similar natural objects (such as bare soil and exposed rock) remains a challenge. Image scenes acquired over natural landscapes are often characterised by an inherently high complexity due to the multitude of land-use types and objects present and the variability of object appearance in images. Factors having a great influence on this complexity include the number of different objects, the amount of interrelations, and the variability of both. Another crucial aspect to be kept in mind is that both the definition of objects and the complexity of the landscape relate to a particular scale, and changes between scales (i.e. upscaling and downscaling) are not straightforward. Therefore, approaches based on multi-scale assessments are a promising scientific trend for the coming years (Burnett and Blaschke, 2003; Ehlers, 1991; Hoberg et al., 2010). To take account of such factors, techniques like automatised modelling, contextual reasoning, and the internal evaluation of intermediate results (cf. Part I, Chapter 6) have demonstrated their importance over recent decades. It is obvious that such techniques must be integral parts of an extraction system to attain reasonably good results for a range of scenes. More details on these issues are presented in Hinz (2008) and Part I, Chapter 2, where information about current issues and the challenges of automatic object extraction are discussed.

For the motivation of this study, it is especially important that most approaches on automatised building mapping are based on almost perfect data availability. This is relevant especially to issues of spectral and spatial resolution, image overlap, and revisit time of the sensor. As mentioned above, similar data bases are often not available when working in developing countries. Hence, most methods published so far cannot be applied to the study site presented herein, and different strategies on data exploitation have to be developed.

This study is based on several years of research in automatised dwelling detection in Benin. In Sturm et al. (2010) the first approaches to automisation were developed and compared to manual mapping; in Sturm et al. (2011) and Sturm-Hentschel et al. (2013) these approaches were further improved using different features and different mapping algorithms. A null model for future settlement change was developed. This contribution (1) summarises these works, (2) improves mapping algorithms further yielding the highest robustness and accuracy of all approaches developed so far, and (3) validates the null model by including a more recent data set than in previous work.

13.3 Study Area and Data Availability

Within this section, the study area and the processes that occur at the coast of Benin will be introduced. Moreover, we will discuss, which data have been available for the study.

13.3.1 The Coastal Area of Benin

The problems *of* and possibilities *for* settlement change assessment based on automatised settlement process studies will be exemplified for the coastal area of Benin in Western Africa, which can be considered a typical example of (1) problems associated with settlement processes in coastal areas of developing countries and (2) problems associated with applying remote sensing in-data scarce situations (cf. Sturm-Hentschel 2013; Vogt et al. 2007). The coast is a graded shoreline with numerous lagoons and spits, in morphological terms, showing low tidal effects. The focus of this study will be the village of Èkpè, a suburban area close to Cotonou, which is the administrative and economical capital.

The study area, presented in Figure 13.1 is representative of the most densely populated areas of Benin. Èkpè has been experiencing very high settlement dynamics during recent decades. In central Benin, migration processes including their driving forces and implications have been relatively well documented in comparison to other regions of Benin and Western Africa. Nonetheless, there has been rather little research for the coastal line itself, where the intensity and dynamics of settlement processes are highest. Here, risks associated with these processes are expected to be highest (Sturm et al., 2007; Vogt et al., 2007). This contribution is another step towards a sufficiently profound analysis of the intensity and dynamics of these processes as is required for spatial planning.

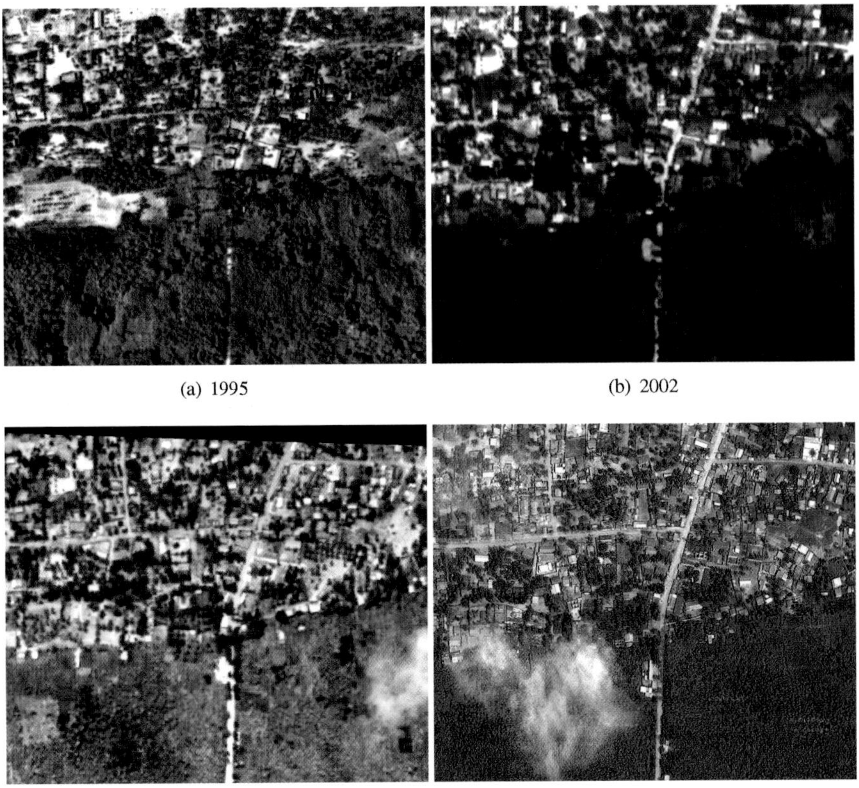

(a) 1995 (b) 2002

(c) 2007 (d) 2013

Figure 13.2 Intensity images of available data: (a) aerial image (1995), (b) QuickBird image (2002), (c) aerial image (2007), (d) aerial image (2013).

13.3.2 Data Availability and Limitations

Despite thorough data availability assessment, it was not possible to find a sensor that would cover the relevant study area (1) at all relevant points in time (2) at sufficiently high spatial resolution to identify individual buildings. Therefore, a multi-sensorial concept had to be developed. This concept is based on a time series that had to be established on a data base of four different images. For a more thorough discussion, cf. Sturm-Hentschel (2013). The first image is a grey-scale aerial image (taken in March 1995 at a scale of 1:30,000). The second is a QuickBird image (acquired in December 2002). The third image is a colour aerial image (acquired in 2007 at scale of 1:20,000), and the most recent image is a colour aerial image (acquired in 2013 at scale 1:20,000) see Figure 13.2. The spatial resolution (ground sampling distance, GSD) of the aerial images (1995, 2007 and 2013) is around 2 m. In contrast, the multi-spectral image acquired by QuickBird has a spatial resolution of 2.4 × 2.4 m, whereas the panchromatic (PAN) channel has a spatial resolution of 0.6 × 0.6 m. The aerial images of 1995, 2002 and 2007 were geo-rectified based on the PAN-QuickBird data set of 2002 and resampled to 0.6 × 0.6 m pixel size. It is obvious that this study represents a typical example of scarce data availability as mentioned above. Automated mapping of individual dwellings will thus have to be adapted to a relatively high heterogeneity in the input data.

13.4 Methods for Settlement Process Studies based on Pattern Analysis

This study develops an approach to perform settlement analysis using remote sensing techniques applied to dynamic and complex landscapes that are represented by poor data bases. At first, building detection is performed using automated methods. Afterwards, the settlement boundary and the settlement density are assessed.

13.4.1 Pattern Analysis: Detection of Individual Dwellings

The approach for automatised building detection developed herein is based on preprocessing of data, segmentation and image computation, multiple classification and accuracy assessment.

13.4.1.1 Preprocessing

The main analysis of the transformation processes occurring at the coast of Benin in the PhD thesis of Sturm-Hentschel (2013) is based on manual building detection. Manual detection (i.e. photointerpretation) is traditionally assumed to produce the most precise maps remote sensing has to offer (Congalton, 1991; Sturm et al., 2010). They provide an excellent database for process identification, risk assessment and vulnerability analysis. However, manual classification of dwellings is time consuming and expensive and may not be feasible for larger areas like the entire coast of Benin. Therefore, automated methods for building detection and process analysis are required (Sturm et al., 2010; Sturm-Hentschel et al., 2013).

Due to the poor availability of data with sufficiently high spatial resolution along the entire coastal range, vulnerability assessments in larger areas will always have to be performed using an ensemble of different data types. In order to process this ensemble with comparable methods, the common denominator of all data sets has to be identified. This is intensity information, which is present in each data type (grey-scale and colour aerial images as well as optical satellite data). For this purpose, each image was IHS (intensity, hue, saturation) transformed and only the intensity information is used for automisation[1]. Subsequently, each image was resampled to match a comparable resolution, which is crucial for process analyses that aim to identify individual buildings.

It may be argued at this point that using only intensity and not multi-spectral information is an unnecessary under-exploitation of available information. We entirely acknowledge this point. However, when transferring to larger regions it has to be decided which of several caveats to accept. Although we acknowledge that using multi-spectral information may increase accuracy, the fact that multi-spectral information is only present for small areas cannot be ignored. Furthermore, as Figure 13.2 shows, buildings in Benin are not so much differentiated by their colour, than by their shape and their contrast to the image context into which they are embedded into. Thus, if a method is found to represent differences in shape and image context, missing multi-spectral information – either because it does not exist, or because it is intentionally not used – may be compensated. For these reasons, under-exploitation of information is accepted in favour of transferability to other regions.

13.4.1.2 Segmentation and Features

As stated above, a method has to be used that represents image context and the shape of objects within the intensity images. In order to do so, new features need to be derived from the intensity values that transform pixel-based information to information valid for larger entities. A region growing segmentation algorithm is applied to the intensity images to transform the pixel space into a segment space. The segment space is thought to consist of two classes of object: house and non-house segments. These two classes of object are to be distinguished and mapped by classification methods (cf. Part I, Chapter 4). However, before this is possible, the differences between house segments and non-house segments have to be modelled by discriminant features. Numerous approaches have been published to compute such features (cf. Weinmann 2013; Weinmann et al. 2013 and Part I, Chapter 3). Within this project, two approaches have been evaluated in previous work. In Sturm et al. (2010), different segment-based features (e.g. shape, area, eccentricity etc.) were evaluated. In Sturm-Hentschel et al. (2013), mathematical morphology profiles were computed. After thorough re-evaluation, the segment-based features approach from Sturm et al. (2010) is used and extended herein, as it proved to be more transferable when applied to various data sets from Benin. For each segment, the features represented in Table 13.2 (at the end of this Chapter) are used.

13.4.1.3 Classification

As mentioned in the introduction, the present contribution is based on previous work. In Sturm et al. (2010), the Bayes quadratic discriminant classifier (QDA) was used and compared to manual classification. Although the results satisfactorily described the overall process of settlement change, they lacked accuracy and robustness on the scale of individual dwellings. In Sturm-Hentschel et al. (2013), the performance of support vector machines (SVMs) and ensemble-based methods was evaluated. Again, automatised methods were not able to achieve accuracy values comparable to manual detection. As shown by Karaca et al. (2012); Roscher et al. (2012a,b); Braun (2013); Braun et al. (2012) and Braun et al. (2014), alternatives exist that were developed from the SVM model, which tend to have similarly high or higher accuracy values in data sets that are noisy and high-dimensional. As Benediktsson et al. (2007) shows, a combination of several of these methods using a multiple classifier system (MCS) may perform better than any individual method. Herein, a relatively straightforward MCS is developed by combining posteriors of SVMs and two alternatives, the relevance vector machine (RVM) and the import vector machine (IVM). In the following, mathematical foundations will be briefly outlined[2].

From a data set $\mathcal{X}$ affected by noise and being high-dimensional due to its large number of features, n training vectors x_i and scalar labels $y_i, i \in [1, \ldots, n]$ are chosen. On this information, the SVM and its alternatives construct a feature space $\mathcal{H}$ by using some kernel function $K(x_i, x_j) = \langle \phi(x_i), \phi(x_j) \rangle$. Since $\phi(\mathcal{X})$ is a non-linear function of $\mathcal{X}$, which is replaced by the kernel K, $\mathcal{H}$ is a non-linear feature space. For thorough details, see Braun (2013). Within this non-linear space $\mathcal{H}$, a separating plane is defined that assigns data points x_k classes y_k and thus performs classification.

In scientific scenarios that monitor socio-economic processes, ground-labelled training data are usually scarce. Therefore, the classification model should be dependent only on a small subset $\mathcal{V} \subset \mathcal{X}$ of training samples. Therefore, each training data point is attributed to a linear weight ω_i. ω_i controls how much this point influences the separating plane that performs mapping. To optimise the separating plane and make it dependent only on a few data points, ω_i are positive only for the mentioned subset $\mathcal{X}$. For all other points, ω_i equals zero. Finally, ω_0 is introduced, and a bias parameter, which is needed to project the separating plane from the origin of $\mathcal{H}$ to the domain where the actual data are situated within $\mathcal{H}$. Thus, a generalised linear model is defined:

$$y_k = f(x_k; \omega_i) = \sum_{i=1}^{n} \omega_i K(x_k, x_i) + \omega_0 \tag{13.1}$$

Each method – SVM, RVM and IVM – parametrises this model by defining *which* of the data points in $\mathcal{X}$ are used as input points x_i. Such points are named support vectors (SVs), relevant vectors (RVs) or import vectors (IVs), respectively. As shown in Braun et al. (2012) the three methods choose different kinds of points. The SVM chooses borderline points, the points lying closest to the respective other classes in the data set. The RVM chooses only the most prototypical examples and the IVM uses some typical points from the entire distribution. As all methods are sufficiently similar (since they are derived from the same model) to be combinable, but produce decision boundaries which depend on other points in $\mathcal{X}$, a combination of the three methods promises to produce a decision boundary that outperforms each individual method.

Such a decision boundary is derived from an MCS, produced by a very simple rule: the product combiner. This MCS exploits the posteriors of each of the methods, that is, $P_{SVM}(y_k|x_k)$, $P_{RVM}(y_k|x_k)$ and $P_{IVM}(y_k|x_k)$. The joint decision, represented by the MCS is produced as[3]:

$$D_{MCS}(y_k|x_k) = P_{SVM} \times P_{RVM} \times P_{IVM} \tag{13.2}$$

Regarding the urban sprawl process that is relevant herein, it can be seen that (1) Èkpè has considerably grown and (2) its boundary has shifted towards the south and into the wetland (covered by fields and forest) found there. Given the rather short period of just 18 years, the high spatial dynamics of settlement processes become obvious. Urban sprawl thus leads to replacement processes that create pressure on native forest ecosystems. Based on the delineation results, it is possible to construct a null model of urban sprawl. The null model is designed to predict future settlement processes based on the urban sprawl observed. The model is data-driven and basically extrapolates the shift of the settlement boundary in a linear fashion until today (cf. Sturm-Hentschel et al. 2013). Certainly, a data-driven model that employs a linear extrapolation follows assumptions that are overly simplified. There is no reason to assume that the processes that lead to past settlement change will continue to produces similar changes in the future. A more reasonable approach would be to analyse the socio-economic factors that are *responsible* for the observed changes and then use some surrogate data for these socio-economic factors to develop a process-based model. In addition to that, geographical features, land cover (like wet land) and land-use types influence future developments. However, the most crucial point besides identifying relevant socio-economic factors is the availability of spatially explicit surrogate data. Since such surrogate data are not available, a process model cannot be developed at this moment. Hence, herein, a data-driven model is developed and considered as a null model for future investigations which may then collect surrogate data and compare process-based and data-driven models.

The model was developed for three points in time – 1995, 2002 and 2007 – in Sturm-Hentschel et al. (2013). At the time this study was produced, no data for 2013 were available. For the study presented herein, the 2013 data set can be used to validate the model. This is done by (1) computing a model prediction for the settlement boundary in 2013 based on the linear model using the data from 1995 to 2007 and (2) delineating the actual settlement boundary using SVDD on the data of 2013. Figure 13.4 shows the results for observed and modelled settlement boundaries in 2013. The right image shows the settlement boundary delineated from building detection, while the left side shows the prediction of the null model, which did *not* exploit detected dwellings in 2013, but used the extrapolation of the boundaries between 1995 and 2007. As a comparison of the two results shows, the data-driven model produces a reasonable estimate of the boundary actually observed in 2013. However, it has to be acknowledged that the change between 2007 and 2013 has been less pronounced than the changes between 1995 and 2007. In particular, settlements spread far less into the wetland (forest) than predicted by the model.

13.5.3 Results of Settlement Densification

Based on the detected individual dwellings, settlement density is computed for each point in time. As Figure 13.5 shows, settlement density continuously

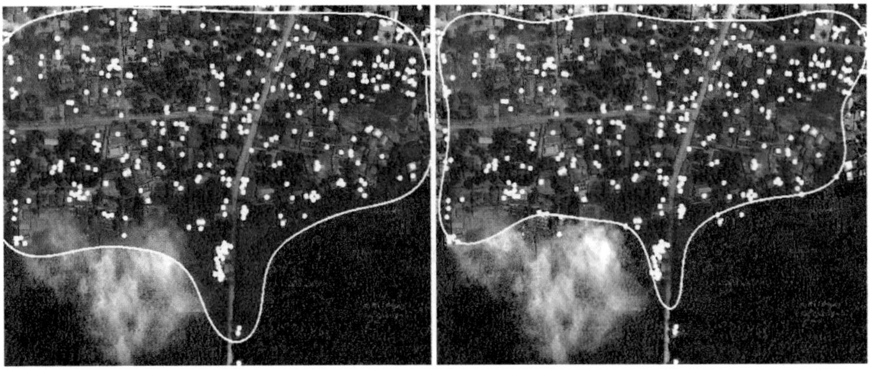

(a) Model.

(b) Observation.

Figure 13.4 Comparison between modelled and observed settlement boundary in 2013 (classified dwellings marked as dots). (a) Linear extrapolation model based on the data from 1995 to 2007. (b) Observed settlement boundary based on the data of 2013.

increased between 1995 and 2013. Again, as for the shift in the settlement boundary, the major part of this change took place between 1995 and 2007. Furthermore, the trend towards an increasing density along the southern road that crosses the wetland towards the coast is lower. Although some settlements were established there, and further establishment may be expected, the presence of settlements is not as dense as in the northern part. Nonetheless, it can be concluded that besides a shift of the settlement boundary, a considerable densification *within* existing settlements occurred.

13.6 Discussion

The previous sections presented a proposition on how to approach situations where small-scale spatial information on settlements has to be analysed using imperfect data availability. The main objectives are pattern and process analyses in risk scenarios related to urban sprawl in developing countries.

The first step was a pattern analysis that aimed to map individual dwellings within the intensity images. The features used were all feature related to the shape of objects, and no spectral information was used. Three different classification methods were joined in an MCS. Clearly, the accuracy values of building detection are not fully satisfactory. None of the individual classification methods (SVM, RVM and IVM) nor the MCS provided accuracy values close to 90%, a value that is sometimes considered as a yardstick value for further processing (Shao and Wu 2008). Every result was considerably affected by both false positives and false negatives. However, it should be kept in mind, that the mapping algorithm developed can be readily applied to the data sets of three different sensors. Furthermore, because intensity data can be derived from almost any remote sensor, the algorithm is in principle also applicable to data not evaluated herein, as long as the spatial resolution of the intensity data is comparable.

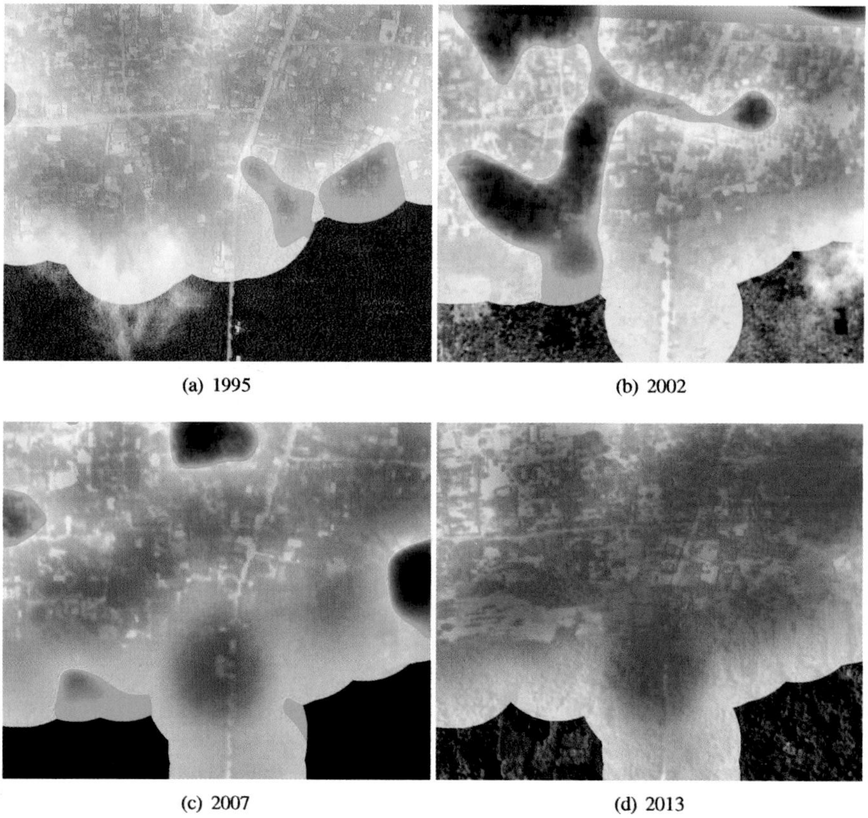

Figure 13.5 Quadratic kernel estimation of settlement density.

Lastly, it was possible to at least describe the overall distribution of houses in Èkpè and to distinguish settled areas from non-settled areas (e.g. the fields and forest in the wetland in the south of the study area). For these reasons, the results are considered sufficiently reliable to be used in process analysis.

The first process analysis was to delineate the settlement boundary in order to analyse its changes. For this purpose, another mapping method – SVDD – was applied to delineate the boundary. SVDD is related to SVM; it represents an absolute classification method, while the latter represents a relative classification method (Part I, Chapter 4). In order to be able to derive a meaningful settlement boundary from an imperfect mapping of individual buildings, certain constraints had to be met. Since some objects were recognised as buildings that in fact do not represent as such, boundary delineation had to be insensitive to outliers. Because SVDD is designed to ignore some data points that are far from the rest of the distribution within the feature space, this criterion was fulfilled. Some objects in the south did not influence the SVDD strongly, so for each point in time, a meaningful boundary was produced. Furthermore, as the method is able to produce convex/concave outlines, the boundary produced well resembled the distribution that a human operator would draw. Thus, from an imperfect mapping of individual dwellings, it is possible to derive the distribution of urbanisations without the requirement of human fieldwork or visual digitisation.

In previous work, a null model was designed that extrapolated past changes (between 1995 and 2007) into the future. When the model was published, no validation data were available. This situation changed with the acquisition of the 2013 image. The model was validated against the new data set. The boundary of the settlement predicted by the model for 2013 was compared to the actual boundary in 2013. In general terms, the model produced reasonable results. The major changes predicted between 2007 and 2013 compare relatively well to the actual situation. However, the model overrated the general shift of the boundary into the (wooded) wetland and underrated the shift along the road. Here, clearly, the limits of data-driven modelling become obvious. The model employed is not designed to recognise a change of the process when reaching the wetland, respectively forest frontier. Clearly, the obstacle of having to cut down trees prevents people from settling within the forests, especially when combined with wet ground. Instead, people prefer to use sites that have already been opened by other processes, such as fields or the construction of the road into the south. The data-driven model is incapable of capturing these circumstances. Therefore, it overrates change in some parts and underrates change in others. However, again, it has to be kept in mind that data for more sophisticated, process-oriented models are lacking. Therefore, in order to give political stakeholders at least some general orientation, it is helpful to develop null models that extrapolate past changes into the near future. Maps derived from such models for larger areas (e.g. the entire coast line) may inform spatial planners, at least, regarding where past situations have to be expected to aggravate and where not, and may distinguish between highly dynamic and less dynamic areas.

A second process analysis concerned settlement densification. Here, kernel density estimation was applied. Kernel density estimation is related to kernel-based classification (Part I, Chapter 4). The main difference is that kernel density estimation uses *all* data point to estimate the distribution, while kernel-based classification uses only a subset (SVs, RVs, or IVs respectively). Here, all data points (dwellings) had to be taken into account. Kernel density estimation represents a statistical standard technique for estimating probability distributions. Yet, it was found to provide convincing estimates of settlement density which can be derived from individual buildings. The results indicate a process of relatively continuous densification of Èkpè – a trend that is typical for the coastal area of Benin.

Hence, two processes have been identified: a shift of the settlement boundary and an intensification of building density. Both processes were derived from a poor data base applying a similar mapping technique to heterogeneous data. Since the method is fully automatised and required only a few labelled houses in situ, it is ready to be applied to larger study sites along the coast of Benin.

It provides estimates for areas where the density of buildings has increased and where settlements spread. Both processes need to be monitored, as they can create land-use pressures and conflicts with other land-use types (Sturm-Hentschel, 2013).

13.7 Conclusion and Outlook

When it comes to the analysis of processes reaching back long before today's 'Sentinel Era', scarce data availability is a frequent issue when integrating remote sensing data into the analysis of patterns and processes in risk-oriented research scenarios in developing countries. In these countries, settlement processes are often more dynamic than in developed countries. To assess vulnerability of households, individual dwellings have to be detected. While the assessment of urbanisation processes using remote sensing is well established for Europa, Australia and the USA, empirical knowledge and expertise to tackle similar situations in developing countries is narrow. Thus, in developing countries scenarios, analysts face the challenge to address most complex analytical requirements with least established data bases.

This chapter has presented one approach to address such issues. A very heterogeneous and incomplete set of images was transformed to a more homogeneous data basis – the intensity images. Within these images, the objects of interest (i.e. individual dwellings) were described by shape features. Then, the settlement of Èkpè was mapped using modern pattern recognition techniques. Although despite several years of research, mapping of individual dwellings could not be achieved with the same level of accuracy as for European cities, for example, the general distribution of the settlement could be mapped with sufficiently high accuracy to be used in subsequent analyses.

The analyses that followed were process analyses that assessed the shift of the settlement boundary and densification of the settlement. Again, state-of-the-art pattern recognition techniques were applied. Both processes could be analysed with convincing results. SVDD successfully provided a representation of the settlement boundary, while kernel density estimation was applied to show how the number of dwellings per area in Èkpè continuously increased. Furthermore, the linear extrapolation model developed in earlier studies was validated using a more recent data set. It provided reasonable results although it has to be assumed that settlement dynamics will change once the boundary of Èkpè entirely reaches the wetlands, respectively forests. Although it is acknowledged that linear extrapolation is only valid under the assumption that the drivers of settlement change remain constant, these result demonstrate the general applicability of the analysis herein for modelling future changes. Methodologies for modelling future settlement changes are especially important in developing countries. In developed countries, spatial planning can generally be based on the assumption that illegal building constructions are rather irrelevant and that people will mostly construct buildings under legal supervision and in designated

Table 13.2 Features used for mapping of individual dwellings. Note that the descriptions for features 1–12 are taken from the documentation of Matlab R2013a, which was used via the command regionprops to compute those features.

No.	Name	Description
1	Area	Scalar; the actual number of pixels in the region. (This value might differ slightly from the value returned by bwarea, which weights different patterns of pixels differently.)
2	Maj. Axis Length	Scalar specifying the length (in pixels) of the major axis of the ellipse that has the same normalised second central moments as the region. This property is supported only for 2-D input label matrices.
3	Min. Axis Length	Scalar; the length (in pixels) of the minor axis of the ellipse that has the same normalised second central moments as the region.
4	Eccentricity	Scalar that specifies the eccentricity of the ellipse that has the same second-moments as the region. Ratio of the distance between the foci of the ellipse and its major axis length.
5	Orientation	Scalar; the angle (in degrees, ranging from -90 to 90Â°) between the x-axis and the major axis of the ellipse that has the same second-moments as the region.
6	ConvexArea	Scalar that specifies the number of pixels in ConvexImage: Binary image (logical) that specifies the convex hull, with all pixels within the hull filled in (i.e., set to one). The image is the size of the bounding box of the region.
7	Euler Number	Scalar that specifies the number of objects in the region minus the number of holes in those objects. This property is supported only for 2-D input label matrices (8-connectivity).
8	Equiv. Diameter	Scalar that specifies the diameter of a circle with the same area as the region. Computed as sqrt(4*Area/pi).
9	Solidity	Scalar specifying the proportion of the pixels in the convex hull that are also in the region. Computed as Area/ConvexArea.
10	Extent	Scalar that specifies the ratio of pixels in the region to pixels in the total bounding box. Computed as the Area divided by the area of the bounding box.
11	Perimeter	Scalar; the distance around the boundary of the region (distance between each adjoining pair of pixels around the border of the region).
12	FilledArea	Scalar specifying the number of pixels in FilledImage.
13	Ratio P/A	Perimeter/Area: The relative portion of pixels that are on the margin
14	Ratio MiAL/MaAL	Minor Axis Length/Major Axis Length: Shape feature of the ellipse defined by the segment
15	Ratio Erod/Segm. 1	Number of Pixels in Image Eroded with Disk, size 1/Number of Pixels in Segment
16	Ratio Erod/Segm. 2	Number of Pixels in Image Eroded with Disk, size 2/Number of Pixels in Segment
17	Ratio Dilate/Segm. 1	Number of Pixels in Segment/Number of Pixels in Image Eroded with Disk, size 1
18	Ratio Dilate/Segm. 2	Number of Pixels in Segment/Number of Pixels in Image Eroded with Disk, size 2

housing estates. The same is not true in developing countries. There, the government frequently has no detailed information about ongoing processes unless they are assessed in the field. This is especially worrisome when vulnerability issues are being addressed. Highly dynamic coastal environments, for example due to their vulnerability to global change, require especially reliable forecasts in spatial planning to avoid an exposure of people to risks (e.g. changes in the coastline or sea level rise). This chapter has discussed how to approach such situations and meet the demands of spatial planners. Nonetheless, despite the generally convincing results of this chapter, much more research is required to make remote sensing fully compatible to situations such as the one given in Èkpè.

References

Adger, N. (1999). Social vulnerability to climate change and extremes in coastal Vietnam. *World Development*, 27(2):249–269.

Benediktsson, J., Chanussot, J., and Fauvel, M. (2007). Multiple classifier systems in remote sensing: from basics to recent developments. In *Multiple Classifier Systems*, pages 501–512. Springer, New York, USA.

Braun, A. (2010). Evaluation of one-class SVM for pixel-based and segment-based classification in remote sensing. *International Archives of the Photogrammetry, Remote Sensing and Spatial Information Sciences*, 38:160–165.

Braun, A. (2013). *Eine fernerkundungsgestützte geoökologische Prozessanalyse zum Risikozusammenhang zwischen Landnutzung und Biodiversität an einem Beispiel aus Chile*. Ph.D. thesis, Karlsruhe Institute of Technology (KIT), Karlsruhe, Germany.

Braun, A., Rojas, C., Echeverria, C., Rottensteiner, F., Bahr, H., Niemeyer, J., Aguayo, M., Kosov, S., Hinz, S., and Weidner, U. (2014). Design of a spectral–spatial pattern recognition framework for risk assessments using landsat data – A case study in Chile. *IEEE Journal of Selected Topics in Applied Earth Observations and Remote Sensing*, 7(3):917–928.

Braun, A., Weidner, U., and Hinz, S. (2012). Classification in high-dimensional feature spaces – Assessment using SVM, IVM and RVM with focus on simulated EnMAP data. *IEEE Journal of Selected Topics in Applied Earth Observations and Remote Sensing*, 5(2):436–443.

Burak, S., Dogan, E., and Gazioglu, C. (2004). Impact of urbanization and tourism on coastal environment. *Ocean & Coastal Management*, 47(9):515–527.

Burges, C. (1998). A tutorial on support vector machines for pattern recognition. *Data Mining and Knowledge Discovery*, 2(2):121–167.

Burnett, C. and Blaschke, T. (2003). A multi-scale segmentation/object relationship modelling methodology for landscape analysis. *Ecological Modelling*, 168(3):233–249.

Capobianco, M., DeVriend, H.and Nicholls, R., and Stive, M. (1999). Coastal area impact and vulnerability assessment: the point of view of a morphodynamic modeller. *Journal of Coastal Research*, 12(2):701–716.

Cicin-Sain, B. and Knecht, R. (1998). *Integrated coastal and ocean management: concepts and practices*. Island Press, Washington D.C., USA.

Clark, G., Moser, S., Ratick, S., Dow, K., Meyer, W., Emani, S., and Schwarz, H. (1998). Assessing the vulnerability of coastal communities to extreme storms: the case of Revere, MA., USA. *Mitigation and Adaptation Strategies for Global Change*, 3(1):59–82.

Congalton, R. (1991). A review of assessing the accuracy of classifications of remotely sensed data. *Remote Sensing of Environment*, 37(1):35–46.

Dennis, R., Mayer, J., Applegate, G., Chokkalingam, U., Colfer, C., Kurniawan, I., Lachowski, H., Maus, P., Permana, R., and Ruchiat, Y. (2005). Fire, people and pixels: linking social science and remote sensing to understand underlying causes and impacts of fires in Indonesia. *Human Ecology*, 33(4):465–504.

Ehlers, M. (1991). Multisensor image fusion techniques in remote sensing. *ISPRS Journal of Photogrammetry and Remote Sensing*, 46(1):19–30.

Epstein, J., Payne, K., and Kramer, E. (2002). Techniques for mapping suburban sprawl. *Photogrammetric engineering & remote sensing*, 63(9):913–918.

Gundlach, E. and Hayes, M. (1978). Vulnerability of coastal environments to oil spill impacts. *Marine technology society Journal*, 12(4):18–27.

Hinz, S. (2008). Automatic object extraction for change detection and GIS update. *International archives of photogrammetry, remote sensing and spatial geoinformation sciences*, 37(B4):277–284.

Hoberg, T., Rottensteiner, F., and Heipke, C. (2010). Classification of multitemporal remote sensing data using conditional random fields. In *2010 IAPR Workshop on Pattern Recognition in Remote Sensing (PRRS)*, pages 1–4. IEEE, New York, USA.

IPCC (1996). *Climate Change 1995: Economic and Social Dimensions of Climate Change - Contribution of the Working Group III to the second Assessment Report of the Intergovernmental Panel on Climate Change (IPCC).* Cambridge University Press, Cambridge UK.

Jat, M., Garg, P., and Khare, D. (2008). Modelling of urban growth using spatial analysis techniques: a case study of Ajmer City (India). *International Journal of Remote Sensing*, 29(2):543–567.

Karaca, A., Erturk, A., Gullu, M., and Erturk, S. (2012). Comparative evaluation of vector machine based hyperspectral classification methods. In *2012 IEEE International Geoscience and Remote Sensing Symposium (IGARSS)*, pages 4970–4973. IEEE.

Lambin, E., Geist, H., Reynolds, J., and Stafford-Smith, D. (2009). Coupled human-environment system approaches to desertification: Linking people to pixels. In Roder, A. and Hill, J., editors, *Recent advances in remote sensing and geoinformation processing for land degradation assessment.* CRC Press, London, UK.

Lee, S., Dunn, R., Young, R., Connolly, R., Dale, P., Dehayr, R., Lemckert, C., McKinnon, S., Powell, B., and Teasdale, P. (2006). Impact of urbanization on coastal wetland structure and function. *Austral Ecology*, 31(2):149–163.

Liverman, D. and Cuesta, R. (2008). Human interactions with the Earth system: people and pixels revisited. *Earth Surface Processes and Landforms*, 33(9):1458–1471.

Liverman, D., Moran, E., Rindfuss, R., and Stern, P. (1998). *People and pixels: linking remote sensing and social science.* National Academy Press, New York, USA.

McGranahan, G., Balk, D., and Anderson, B. (2007). The rising tide: assessing the risks of climate change and human settlements in low elevation coastal zones. *Environment and Urbanization*, 19(1):17–37.

Platt, J. (1999). Probabilistic outputs for support vector machines and comparisons to regularized likelihood methods. *Advances in large margin classifiers*, 10(3): 61–74.

Roscher, R., Förstner, W., and Waske, B. (2012a). I^2vm: Incremental import vector machines. *Image and Vision Computing*, 30(4):263–278.

Roscher, R., Waske, B., and Förstner, W. (2012b). Incremental import vector machines for classifying hyperspectral data. *IEEE Transactions on Geoscience and Remote Sensing*, 50(9):3463–3473.

Scheren, P., Kroeze, C., Janssen, F., Hordijk, L., and Ptasinski, K. (2004). Integrated water pollution assessment of the Ebrié Lagoon, Ivory Coast, West Africa. *Journal of Marine Systems*, 44(1):1–17.

Shao, G. and Wu, J. (2008). On the accuracy o landscape pattern analysis using remote sensing data. *Landscape Ecology*, 23(5):505–511.

Small, C. and Nicholls, R. (2003). A global analysis of human settlement in coastal zones. *Journal of Coastal Research*, 19(3):584–599.

Sturm, U., Braun, A., Hinz, S., and Vogt, J. (2010). Settlement process studies in developing countries using diverse remote sensing data types. In Lenz-Wiedemann, V. and Bareth, G., editors, *Kölner Geographische Arbeiten 92*. Universität zu Köln, Köln, Germany.

Sturm, U., Braun, A., Hinz, S., and Vogt, J. (2011). Characterizing settlement changes and its implications in the coastal area of Benin using high resolution data. In *3rd EOGC – Earth Observation on Global Changes*.

Sturm, U., Teka, O., Vogt, J., Bähr, H., and Buttschardt, T. (2007). Analyse von Prozessen im Küstenraum Benins – Grundlage für planerische Steuerung. In Gönnert, G., Pflüger, B., and Bremer, J., editors, *Geographie der Meere und Küsten*. Coastline Reports 9.

Sturm, U. and Weidner, U. (2009). Further investigations on segmentation quality assessment for remote sensing applications. In *ISPRS Hannover Workshop 2009 High-Resolution Earth Imaging for Geospatial Information*.

Sturm-Hentschel, U. (2013). *Fernerkundungsgestützte Prozessanalyse im Küstenraum Benins*. Ph.D. thesis, Karlsruhe Institute of Technology (KIT), Karlsruhe, Germany.

Sturm-Hentschel, U., Braun, A., Hinz, S., and Vogt, J. (2013). Detecting and characterizing settlement changes in developing countries using VHSR data: Case of the coastal area in Benin. In Krisp, J., Meng, L., Pail, R., and Stilla, U., editors, *Earth Observations of Global Changes*. Springer LNG& C, Springer, Heidelberg, Germany.

Tax, D. and Duin, R. (1999). Support vector domain description. *Pattern recognition letters*, 20(11):1191–1199.

Tipping, M. (2001). Sparse bayesian learning and the relevance vector machine. *The journal of machine learning research*, 1:211–244.

Turner, R., Subak, S., and Adger, W. (1996). Pressures, trends, and impacts in coastal zones: interactions between socioeconomic and natural systems. *Environmental management*, 20(2):159–173.

Vogt, J., Sturm, U., Teka, O., Bähr, H., and Sinsin, B. (2007). Résultats préliminaires de l'analyse évolutive de la zone côtière du Benin: Cas spécifiques des Arrondissements de Avlékété et de Sèmè. In *Actes du 1er colloque de l'UAC des Sciences, Cultures et Technologies, Géographie. Section Géographie*, pages 419–428.

Weinmann, M. (2013). Visual features: From early concepts to modern computer vision. In Farinella, G., Battiato, S., and Cipolla, R., editors, *Advanced Topics in Computer Vision*, pages 1–34. Springer, London, UK.

Weinmann, M., Jutzi, B., and Mallet, C. (2013). Feature relevance assessment for the semantic interpretation of 3D point cloud data. In *ISPRS Annals of the Photogrammetry, Remote Sensing and Spatial Information Sciences, Vol. II-5/W2.*

Zhu, J. and Hastie, T. (2005). Kernel logistic regression and the import vector machine. *Journal of Computational and Graphical Statistics*, 14(1):11–21.

Notes

1. For QuickBird, the following channels were used for IHS transformation: red=channel 3; green=channel 2; blue=channel 1.
2. For further reading, the interested author is referred to Burges (1998) for SVM, Tipping (2001) for RVM and Zhu and Hastie (2005) for IVM.
3. For probabilistic outputs for SVM, that is, $P_{SVM}(y_k|x_k)$, cf. Platt (1999). Note that $D_{IVM}(y_k|x_k)$ does *not* represent posteriors anymore, since the first Komogorov axiom is violated after multiplication. This does not, however, influence the robustness on the decision on x_k.
4. It may be argued at this point, that larger dwellings might be inhabited by larger families and hence, building size *is* a relevant surrogate for population growth. However, there is no evidence to generally assume this relation to hold. Larger buildings can be dwellings of wealthier citizens without being inhabited by more persons, or can be of other use.

Part III
Conclusion

Benchmarking: A Basic Requirement for Effective Performance Evaluation

CHAPTER 14

Martin Weinmann, Andreas Ch. Braun

14.1 Introduction

For many applications it is of great importance that a statement about the performance of proposed approaches can be made, and it is therefore desirable to compare the results obtained with a respective approach to some kind of reference. This comparison in turn allows us to judge about the suitability of an approach, and a further comparison of different approaches with respect to the reference allows us to draw conclusions about which approach is best suited best to solve a particular problem. While the comparison to a reference has of course been part of a huge number of investigations, most of the involved data is not shared for other people to test their own approaches on the same data. This is particularly meaningful if one takes into account that the results of any approach may vary significantly from one data set to another. Accordingly, it is desirable to have publicly available benchmark data sets on which the performance of approaches addressing a certain problem can be evaluated. In the last decade, the number of publicly available benchmark data sets has been drastically increased, addressing novel research domains and partially representing more challenging data sets. Among others, a diversity of benchmark data sets has recently been published for multi-platform photogrammetry (Nex et al., 2015), mobile robotics and autonomous driving research (Geiger et al., 2013), egomotion estimation by involving laser scanning data and/or multi-fisheye camera imagery (Urban and Jutzi, 2016), 3D scene analysis (Hackel et al., 2016; Munoz et al., 2009; Paparoditis et al., 2014; Rottensteiner et al., 2012; Serna et al., 2014), classification of hyperspectral imagery (Braun et al., 2015), material classification (Sharan et al., 2009; Weinmann et al., 2014), object recognition and image classification (Deng et al., 2009; Lin et al., 2014; Russakovsky et al., 2015), or object detection and tracking (Schmidt and Hinz, 2011). To give the reader an impression about such benchmark data sets, we focus on selected data sets addressing three different categories.

14.2 3D Reconstruction, Image Orientation and Navigation

The *ISPRS Benchmark for Multi-Platform Photogrammetry* (Nex et al., 2015) focuses on the use of state-of-the-art multi-sensor data (oblique airborne imagery, unmanned aerial vehicle (UAV)-based imagery and terrestrial imagery) acquired over an urban area, and it aims to evaluate the performance of different approaches for dense image matching and image orientation.

Regarding the task of dense image matching in order to obtain a 3D reconstruction of a considered scene, one scenario is focused on dense image matching in oblique and nadir airborne imagery where larger-scale differences and occlusions have to be addressed, while another scenario is focused on dense image matching in a combined terrestrial and UAV image block. The latter scenario thus not only involves nadir and oblique imagery but also ground-based horizontally oriented imagery, which, in turn, makes the task of dense image matching even more challenging.

Regarding the task of image orientation, the main focus is on the image orientation assessment of different image block configurations. This task comprises four different scenarios: 1) orientation of large airborne multi-view image blocks, 2) orientation of large airborne multi-view image blocks including UAV imagery, 3) image orientation and calibration across UAV and terrestrial platforms, and 4) absolute image orientation of UAV images using airborne nadir and oblique imagery and their known exterior and interior orientation.

The evaluation of both dense image matching and image orientation relies on ground truth data derived from terrestrial laser scanning data, aerial laser scanning data, topographic networks and global navigation satellite system points. For a detailed description, see (Nex et al., 2015).

14.3 Object Detection and Tracking

The *KIT AIS data set* (Schmidt and Hinz, 2011) has been presented as a benchmark data set for detecting and tracking people or cars in aerial image sequences. The data set consists of several labelled aerial image sequences, where reference positions in the form of image coordinates are provided for people or cars. These reference positions have been obtained by manually marking the tracks of all persons or cars, respectively. For the aerial image sequences addressing people tracking, the single images contain up to several hundreds of individuals per frame. Exemplary images and the respective labelling are shown in Figure 14.1.

14.4 Object Classification and Semantic Segmentation

A variety of approaches in remote sensing focuses on the classification of hyperspectral imagery, for which one of the more recent data sets is represented by the *EnMAP benchmark data set* (Braun et al., 2015). This data set is based on the simulated EnMAP Alpine Foreland data set (Guanter et al., 2009; Segl et al., 2010) consisting of an image of 1000 × 1000 pixels that corresponds to an area of about 30 km × 30 km around the Ammersee in Bavaria, Germany. A near natural colour visualisation of the image is provided in Figure 14.2 and reveals a diversity of water, vegetation, agricultural and urban/industrial classes. For

Figure 14.1 Example images of the KIT AIS data set (Schmidt and Hinz, 2011) and labelling.

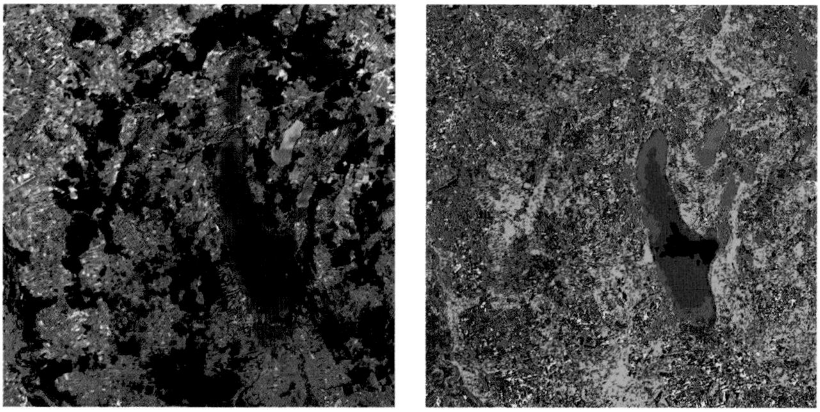

Figure 14.2 The EnMAP benchmark data set: simulated hyperspectral data covering 244 bands are available for each pixel of the scene (left image courtesy of Dr K. Segl), and the goal consists in assigning respective class labels (right).

each pixel, information on 244 simulated spectral channels covering the spectral range from 420 nm to 2450 nm with a varying spectral sampling of 6.5–10 nm is available. In comparison to the simulated EnMAP Alpine Foreland data set, the EnMAP benchmark data set also contains a reference labelling for selected pixels, in which 20 different land-cover and land use classes are considered.

Table 14.1 Information about the 20 classes of the EnMAP benchmark data set. Class numbers, names, codes, numbers of training l_x and evaluation pixels l_y and colourmap.

No.	ClassName	ClassCode	$n(l_x)$	$n(l_y)$	Cmap (R,G,B)
1	Water 1	1WAT	140	60	(000,000,040)
2	Water 2	2WAT	139	60	(002,000,122)
3	Water 3	3WAT	140	60	(003,000,237)
4	Water 4	4WAT	140	60	(109,105,255)
5	Water 5	5WAT	140	60	(000,066,053)
6	Vegetation 1	1VEG	140	60	(001,071,000)
7	Vegetation 2	2VEG	140	60	(000,185,000)
8	Vegetation 3	3VEG	140	60	(019,255,019)
9	Urban / Industry 1	1URB	140	60	(193,010,000)
10	Vegetation 4	4VEG	140	60	(034,055,000)
11	Agriculture 1	1AGR	140	60	(104,125,000)
12	Agriculture 2	2AGR	140	60	(147,255,101)
13	Urban / Industry 2	2URB	140	60	(219,000,118)
14	Urban / Industry 3	3URB	140	60	(165,165,165)
15	Vegetation 5	5VEG	140	60	(254,254,100)
16	Vegetation 6	6VEG	140	60	(051,010,000)
17	Agriculture 3	3AGR	140	60	(089,033,000)
18	Agriculture 4	4AGR	110	48	(123,067,000)
19	Vegetation 7	7VEG	98	42	(166,123,000)
20	Vegetation 8	8VEG	30	14	(074,063,000)

These classes comprise five water classes, five forest/meadow classes, five agricultural classes, three urban/industrial classes, and two pasture/fallow classes. The single classes have been defined by an expert via manual annotation based on visual differences in the image (considering several channel combinations) and the individual spectra of respective pixels. In order to create a data set difficult to classify, the expert has defined classes that overlap strongly. Figure 14.4 shows the overlap of the 20 classes of the EnMAP benchmark data set, and shows the number of classes that occupy the spectral space at each position (within an interval of $\mu \pm n\sigma$). It becomes obvious that many positions within the spectral space are within the $\pm 3\sigma$ interval of as many as 14 classes.

In total, the training set contains hyperspectral data for 2617 labelled pixels, whereas the test set contains hyperspectral data for 1124 labelled pixels. The spectra are provided as x_i and y_i for training pixels and evaluation pixels, respectively, and the corresponding labels are found in l_x and l_y. A mandatory colourmap is provided and found in the data set and in Table 14.1. More details about the data set are provided in (Braun et al., 2015) and in the table. As the table reveals, few training data are provided for each class in order to make training

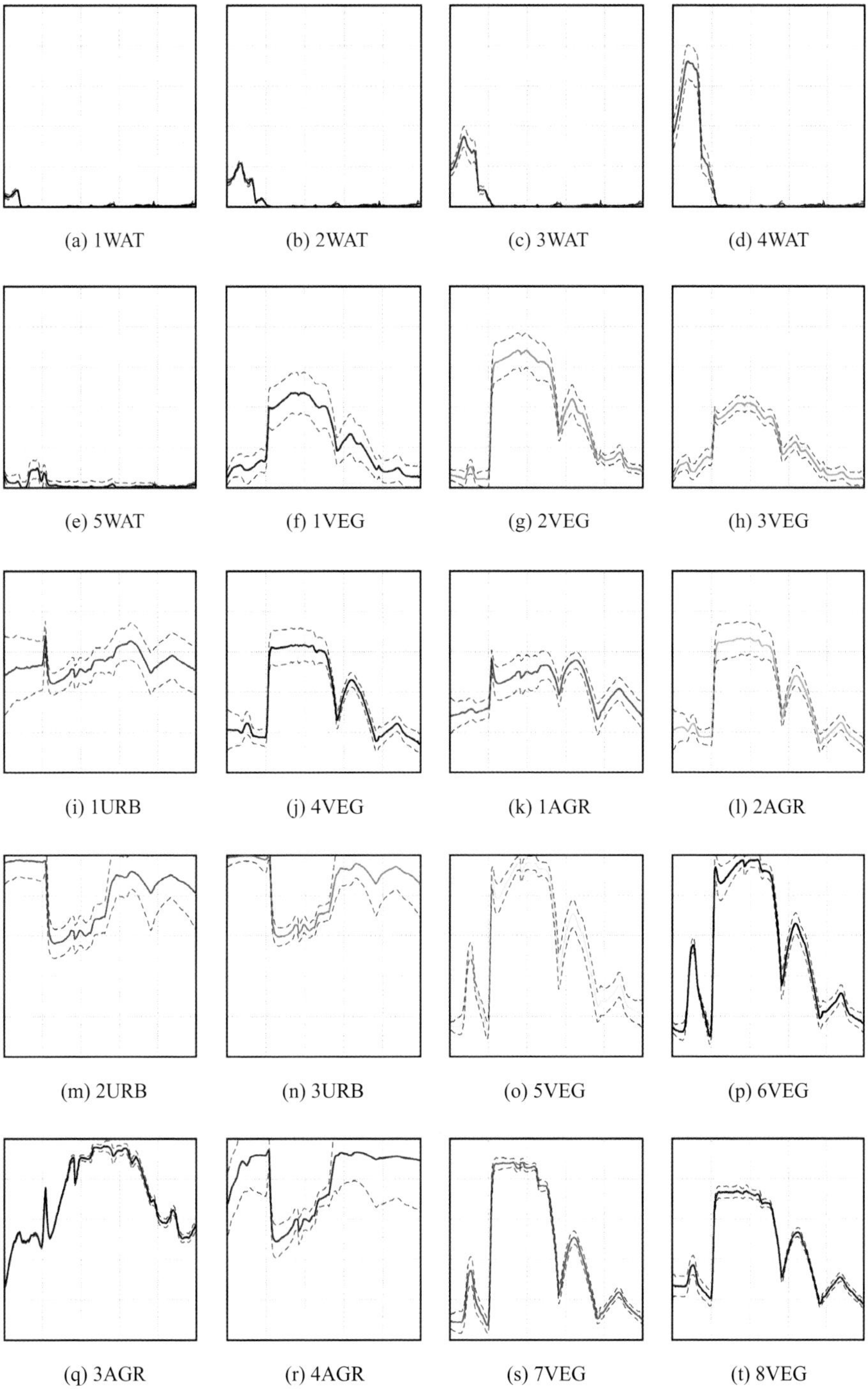

Figure 14.3 The 20 classes provided in the EnMAP benchmark data set. horizontal axis, simulated EnMAP channels; vertical axis, reflectances; spectra shown as mean and standard deviations.

(a) $\mu \pm 1\sigma$ Interval (b) $\mu \pm 2\sigma$ Interval (c) $\mu \pm 3\sigma$ Interval

Figure 14.4 Overlap of the 20 classes of the EnMAP benchmark data set. Shown is the number of classes that occupy the spectral space at each position (within an inverval of $\mu \pm n\sigma$). It becomes obvious that many positions within the spectral space are within the $\pm 3\sigma$ interval of as many as 14 classes.

difficult and simulate field conditions (where seldomly more than around 200 pixels can be labelled in terrain). Furthermore, the numbers of training pixels l_x and evaluation pixels l_y are relatively balanced (with standard deviations of $\sigma(l_x) = 26.25$ pixels and $\sigma(l_y) = 11.00$ pixels). This ensures that all classes are similarly difficult to train in the sense that, for instance, parametric classifiers will be provided the same class of training data to estimate the distribution for each class. Of course, the classes are differently hard to train regarding their spectral features. Furthermore, it ensures that accuracy values, which tend to become biased by disbalances in the evaluation data (such as overall accuracy), provide meaningful figures. Intentionally, no indices to the image positions of labelled pixels are provided. This is also true for us, the makers of the data set. We deleted the indices after extracting the spectral features into x_i and y_i, which is something, of course, we cannot prove. The reason is to avoid manipulations of the results, for example, by assigning falsely labelled pixels the labels of their neighbours in the images.

The instructions are found on the homepage of the editors of this book (Institute of Photogrammetry and Remote Sensing at Karlsruhe Institute of Technology). Basically, they are simple. Train a classifier on x_i and l_x, classify y_i, producing predictions p_y, compute some accuracy values by comparing p_y and l_y. Classify the entire image (provided in the data set), map it using the mandatory colourmap, and submit the results to us for comparison. It should generally be taken into account that some of the classes are spectrally very similar which makes classification more challenging, as Figure 14.3 shows.

To evaluate an approach for classifying 3D point cloud data, one could use the *Vaihingen data set* (Cramer, 2010). This data set comprises an airborne laser scanning data set acquired with a Leica ALS50 system over Vaihingen, a small village in Germany. The acquired data correspond to areas with small multi-storey buildings and many detached buildings surrounded by trees. The

Remote Sensing and Computer Vision Image Analysis: Summary and Recent Trends

CHAPTER 15

Andreas Ch. Braun, Martin Weinmann, Sven Wursthorn, Sina Keller, Andreas Schenk, Boris Jutzi, Stefan Hinz

In 1966, Marvin Minsky at MIT advised an undergraduate student to "*spend the summer linking a camera to a computer and getting the computer to describe what it saw*" (Szeliski, 2010). Forty-two years later, Hanjalic et al. (2008) declared the same subject as "*the holy grail of information retrieval*" when announcing an IEEE special issue. Obviously, the scientific appreciation of image analysis in remote sensing and computer vision has changed tremendously.

From today's perspective, the first statement may seem a fool's errand. Of course, this is by no means true. Image analysis is a process of coding information into images and extracting it from them. With information coded in pictographs, mankind has 14,000 years of experience; with information coded in written characters, we have at least 5000 years of experience. In contrast, our experience with coding and extracting information in *images* is limited to roughly 200 years (Figure 15.1).

In such a limited period, remote sensing and computer vision image analysis developed rapidly, not least because it was engaged to support and promote analyses from a wide range of scientific disciplines. As Chapter 1 in Part I has tried to point out, both remote sensing and computer vision have been applied to countless different scientific applications. Within such a diverse discipline, it is important to have a reliable roadmap at hand to be able to keep pace with the rapid evolution of research. This book has strived to develop such a roadmap. It is structured in theoretical parts, which outline the most important foundations, and application parts, which exemplify how to use them. This work will now be summarised from a global perspective. For the theoretical chapters, this summary focuses on aspects relevant to research. For the application chapters, it emphasises the benefits society gains from remote sensing. Afterwards, recent trends will be indicated to adumbrate where the cutting edge in image analysis may be heading.

15.1 Summary

Features can be thought of as representations of the properties of objects visible in images. Now, consider a really simple example. We have a really simple 2D

Figure 15.1 Three ways of coding and extracting information. Left: Information as pictographs; cave paintings from El Castillo ≈ c.14,000 BP (Valladas et al., 1992), Centre: Information as characters; first human writings from Sumerian Uruk ≈ c.5000 BP (Nissen, 1986), Right: Information as images; first photography, acquired by Joseph Nicéphore Nièpce in 1826 (Smith, 1983).

data set, a 3×3 pixel (10×10 m each) image with only one channel (intensity). We have a lot of prior information on that image. We know that there are only two types of objects, one is a perfectly dark coniferous forest and the second is a perfectly bright road crossing that forest. Our task is merely to *detect* this road. Now note that, despite the absolute simplicity of this example, there are already eight hypotheses where this road could be (along each of the three image rows, along each of the three image columns and along each of the diagonals). For each of these hypotheses, a feature could be defined to detect that road (e.g. the mean intensity along each direction). Logically, when it comes to real-world examples, with thousands of objects, pixels and hypotheses, countless features have been designed and published (apart from automatically learning features from millions of example images). Martin Weinmann, in Part I, Chapter 3, has not merely reviewed, but systematically organised these features. He has developed an ontology helpful to scientists aiming to find the most well suited features to represent the properties of their objects in their images.

The overarching goal of *every* image analysis approach in remote sensing and computer vision is always the same: to make the computer "see" what the human operator wants it to see. Nonetheless, due to the particularities of each individual task, numerous approaches have been developed. Stefan Hinz, in Part I, Chapter 2, has strived to synoptically reintegrate these particular approaches. He has provided an ontology very helpful to image analysis developers that focuses scientists to the *central questions* of all approaches. What kind of information is available for the task? How is this information represented? How does complexity arise within the task, due to properties of the image and the sensor? And how can this complexity be controlled by an appropriate model and an appropriate extraction strategy?

Every student in geodesy learns that *"measurement is worthless without quality assessment"*. Thus, every professor, when teaching even the most introductory lecture, will explain how to produce a confusion matrix and derive some figures like overall accuracy, kappa coefficient, producer and consumer accuracy from it. This may lead to the impression that quality assessment is "an

Figure 15.2 Nike of Samothrace: marble sculpture of around 190 BC, artist unknown, 328 cm in height (Sleeswyk, 1982).

old hat" in remote sensing. Uwe Weidner has shown in Part I, Chapter 6 that this is in no way true. In fact, quality assessment is an intensively studied field of research and, although countless figures and indices have been published, a general framework and mandatory guidelines for quality assessment have been lacking until today.

Images are complex. For instance, a Landsat image with eight channels in 8-bit radiometric resolution offers over $1.8 \cdot 10^{19}$ (256^8) possibilities to represent an object. Some studies, like Braun (2013), consider study sites as large as $13{,}993 \times 11{,}094$ pixels, for which, consequently, there are over $\mathbb{K}_c \approx 2.8 \cdot 10^{27}$ possibilities of how such images may "look" in a continuous image space. It is obvious why these authors strive to *reduce* this complexity by discretizing to just eight land-use classes. In discrete land-use maps, only $\mathbb{K}_d \approx 1.2 \cdot 10^9$ possibilities exist. Image classification can thus be thought of as a task to reduce complexity. It makes the discretised land-use map $2.3 \cdot 10^{18}$ times easier to comprehend for human cognition $(\mathbb{K}_c/\mathbb{K}_d)^1$. In Part I, Chapter 4, Martin Weinmann and Andreas Braun have tried to prepare the reader to understand the application chapters of Part II. They have discussed fundamentals and modern techniques of exemplar-based image classification. The further consideration of contextual information to improve the classification results has been presented by Franz Rottensteiner in Part I, Chapter 5.

What might the face of Nike of Samothrace have looked like? In fact, we could hypothesize this from expert knowledge or - more in the sense of this book - we could hypothesize this by semantic and statistical models derived from representative examples, yet we will never know for sure because her head is missing (Figure 15.2). It was lost at some moment between 190 BC, the year when it was created, and 1863, when the fragments of the statue were recovered (Sleeswyk, 1982). Unfortunately, the original appearance of the statue is nowhere documented. With modern sensors and a methodology like the one provided by Martin Weinmann, Sven Wursthorn and Boris Jutzi in Part II, Chapter 7, a documentation of cultural heritage objects is possible. Of course, documenting artwork is just one of the many applications that the framework allows. As the authors have pointed out, the framework offers a lot of potential for future research, and documenting artwork is merely one example of the many applications that become possible by general research on 3D point cloud processing.

Every car driver is annoyed by traffic jams. At the same time, when *in* a traffic jam, drivers change lane, leave free spaces unused and make the situation worse by trying to overtake others. Stefan Hinz and Jens Leitloff, in Part II, Chapter 8, have shown how remote sensing may help to monitor such situations, to assist the police in preventing traffic jams, and to prompt and help decision makers to plan better roadworks for all drivers.

In democratic countries, at least, people are in general free to move to wherever they want. This societal achievement may have some serious consequences though. In 2010, at a music festival in Duisburg, Germany, 21 young visitors lost their lives and 541 were injured. At a bottleneck point close to the festival's entrance, people from the front started to move backwards, while people from behind continued to try to enter (Blätte and Mamrot, 2010). Panic spread and people got stuck and trampled. Analyses, such as the one by Florian Schmidt and Stefan Hinz in Part II, Chapter 9 can help to prevent such situations. By monitoring the movement of people, protection by restrictions may become unnecessary as dangerous situations become visible in advance.

Land-use analyses are a traditional application of remote sensing. However, as Andreas Braun and Callum Banfield have pointed out in Part II, Chapter 10, even today, reliable maps are not simply taken "off the shelf". These authors mainly show how remote sensing can not only map land-use *classes*, but also land-use change *processes*. Moreover, they emphasise how remote sensing and fieldwork can be applied in an integrative manner, with results from one type of analysis being the nucleus for investigations in the other type of analysis. By doing so, risks for ecosystem services may be predicted even before they become apparent.

For fast acquisition of geospatial data, independent from daytime and weather conditions, the use of synthetic aperture radar (SAR) systems is highly attractive. In Part II, Chapter 11, Antje Thiele, Clémence Dubois and Stefan Hinz have reviewed the fundamentals of such systems and they have presented an exemplary application addressing large-scale mapping and modelling of urban areas. The presented techniques allow conclusions about the number of buildings and an estimation of their main parameters, which is particularly important for supporting urban planning strategies such as monitoring changes or organizing energy supply.

Chaos breaks out when a strong earthquake occurs. In a matter of minutes, infrastructures break down and thousands of persons may be trapped under debris and be at risk of their lives. Miriam Hommel and Thomas Vögtle have shown in Part II, Chapter 12 how laser scanning data can be rapidly processed in an automatised manner to provide the initial information on damage that is crucially required, but which no-one has. If data acquisition and processing are operationally available, such techniques may in fact lead rescue teams to entombed people and help to save their lives, before the critical 72 hours have passed.

In Part II, Chapter 13, Andreas Braun, Ulrike Sturm-Hentschel and Stefan Hinz have shown that remote sensing techniques may be applied to socio-economic analysis. They have especially stressed how such techniques can deal with situations where data scarcity and high process dynamics meet. The most important point in their work is that, although *pattern analysis* may lack accuracy, *process analysis* may still provide valuable insights as long as the reasons for the lack of accuracy are understood and consequences for the analysis are adequately controlled. Such analysis may provide some truth in situations where information from governmental authorities is either lacking or not trustworthy (e.g. due to bad governance).

15.2 The Complexity of Ground Truth Acquisition for Model Validation and Accuracy Assessment

The availability of independent ground truth data is of particular importance for model validation and accuracy assessment. In cases of classification with labelled objects and discrete classes like land-cover and land-use, quality assessment may be achieved by comparison with up-to-date geographic information system (GIS) data or visual inspection of high-resolution satellite data or aerial photography. In these cases, the term "ground truth" is somewhat misleading, as it is mainly a comparison between machine-based classification and human interpretation, with the latter being assumed to be the correct estimate. However, there is a large demand for local and in situ measurements, particularly in the context of modelling and parameter retrieval of so-called biogeophysical parameters, which describe physical, chemical or biological parameters of the atmosphere, biosphere, hydrosphere and the Earth's surface. Often, these parameters cannot be measured directly. While the situation of acquiring reference or ground-truth data might be quite obvious for the case of binary or discrete data (such as objects in the real world), this becomes even harder for the validation of continous parameters derived from remote sensing data. Their combined estimation from remote sensing data and in situ observations is an ongoing field of research and development, with new and improving sensors, methods and algorithms. In the following subsections, we consider two examples in detail.

15.2.1 Example 1: Biochemical Parameters

Biochemical parameters are the underlying quantity in agricultural and ecological and environmental applications of remote sensing. The parameters of primary interest are chemical substances like leaf nitrogen concentration, phosphorus concentration and chlorophyll content, but also geometric parameters like canopy height, leaf size and leaf area index (LAI) (Homolová et al., 2013). Biochemical parameters can be determined using empirical relations between these variables and hyperspectral indices – most prominent vegetation indices like the normalized difference vegetation index (NDVI) and the enhanced vegetation index (EVI). Another approach is the inversion of physical models using radiative transfer functions, which simulate the whole process of light absorption, reflection and canopy scattering (Jacquemoud et al., 2009). A recent trend is the use of Bayesian variable selection and model averaging approaches (Zhao et al., 2013).

Whatever way is chosen to estimate biochemical parameters, a comprehensive and reliable ground truth data set is indispensable to develop and improve retrieval algorithms and to assess their accuracy. Usually, there is not just one and only method to retrieve ground truth data for a specific variable. Instead, several methods based on different chemical or physical principles are available. Moreover, in many cases, these estimates are not directly estimated but indirectly inferred, and thus are also prone to errors and biases.

Collecting ground truth data for plant nitrate concentration and related chlorophyll content is one of several examples of the challenge to retrieve ground truth data. There are at least nine approaches to measure plant nitrogen concentration, each with certain advantages and drawbacks (Muñoz-Huerta et al., 2013). These methods include invasive and destructive approaches, first and foremost tissue analysis based on digestion and combustion, but the use of ion selective electrodes (ISEs), impedance spectroscopy and nitrate test stripes have also been documented as valuable approaches to measure plant nitrate concentration (Azzarello et al., 2012; Goffart et al., 2008). However, for comparison with retrievals from remote sensing, it is most common to estimate plant nitrate concentration using either transmittance and fluorescence on a leaf level (Demotes-Mainard et al., 2008; Goffart et al., 2008; Lin et al., 2010) or active and passive spectroscopy on a canopy level (Goffart et al., 2008; Li et al., 2008; Shaver et al., 2011).

15.2.2 Example 2: Soil Moisture

Soil moisture is an important biogeophysical parameter in the context of agricultural applications. It affects, among other factors, crop health (Hassan-Esfahani et al., 2015). To reduce and optimise the application of fertilisers and pesticides to avoid negative impacts on groundwater resources, the field-scale variability of soil moisture needs to be understood. One key issue is to overcome the spatial variability of soil moisture, ranging from the field scale to the catchment scale and finally to the regional scale. For calibrating, mapping, and characterising spatio-temporal subsurface soil moisture with remote sensing techniques, it is necessary to measure adequate ground truth data at the field side.

Measurements of the soil moisture can be conducted with different sensors: (1) invasive time domain reflectometry (TDR), (2) non-invasive geophysical methods such as ground-penetrating radar (GPR), and (3) non-invasive field spectroradiometer (field spec.). With the aid of each sensor, determined ground truth data, applied for different aspects of soil moisture, are provided.

Since the 1980s, TDR has been the standard method to selectively measure soil moisture within the upper 30 cm of the soil. Manual TDR sampling campaigns in the field provide information about the soil moisture with temporal high resolution for a specific field point and could be coupled with point-by-point field spec. data. The field spectrometer records hyperspectral data in the spectral range 350–2500 nm. Deriving the footprints of soil moisture in the field spectrometer data is possible on the basis of knowledge about the soil moisture process retrieved by the TDR data.

Combining GPR and electrical resistivity tomography (ERT) represents a typical example of deriving information about relevant subsurface structures and properties of the specific soil at the field scale. Based on this data set, information and patterns of subsurface soil moisture are derived. These are then used to analyse the absolute reflectance values of each spectral band in the field spectrometer data. Regarding soil moisture footprints in hyperspectral data, two basic effects are present: (1) dry soil has a higher albedo and (2) dry soil has a higher reflectance than irrigated/wet soil in specific bands (Bryant et al., 2003).

Related patterns are recognised in the field spectrometer data focusing on specific spectral values and bands, correlating with soil water and demonstrating decreasing reflectance values (Babaeian et al., 2015; Ciampalini et al., 2015; Younis and Iqbal, 2015). Since the hyperspectral sensor records reflectance at the surface of the soil, a temporal shift of the soil moisture patterns exists relative to the GPR soil moisture values with the corresponding reflectance values of the hyperspectral data.

Remote sensing technologies such as full-waveform laser scanning are appropriate to draw conclusions about the roughness of soil. This information, combined with the mentioned captured data, functions as additional ground truth data for the hyperspectral field spectrometer data regarding the geophysical classification of the investigated field side.

Among the mentioned techniques, GPR and the hyperspectral field spectrometer measurement deepen the understanding of how soil moisture processes are connected to geophysical and remote sensing signatures on the field (point) side. These data could be regarded as determined ground truth data about soil moisture values and patterns with low spatial, but high temporal resolution. Furthermore, a relationship among the (in situ) measured soil moisture data, the point-based measured field spec. data, the area-based measured GPR data and the data extracted by areal hyperspectral records has to be developed and derived. For example, hyperspectral sensors mounted on a cable car installation or drone provide information about soil moisture within the spectral range over an entire area. With the aid of a settled relationship between all data based on sampled ground truth data of, for instance, GPR, TDR, field spec. and laser scanning, essential information is provided to explore the linkages between remote sensing and soil moisture signatures on an increased spatial resolution to map and analyse soil moisture on a catchment scale.

15.3 Recent Trends from our Perspective

Within this section, we will try to outline some recent trends from our perspective. Any attempt to make a judgement on *recent trends* within a community as large as the one targeted herein runs the risk of being characterised with implicit arrogance. Why is that? First, given the exponential increase in the number of published approaches, what is the percentage of recent publications, we can at least be aware of? Second, what are the criteria to separate relevant from less relevant approaches and who on Earth gives *us* the right to judge on them? And third, what the heck is a *trend*? In order to avoid the pitfall of naive arrogance, it is important to keep in mind that such an attempt is highly incomplete, subjective and biased. We do not claim to establish definitive power over the *hot topic* of our community herein. We are simply trying to outline what has caught *our* attention during recent years. And we trust that discourse and contradiction by our colleagues will help to jointly identify what *truly* can be considered the recent trends.

15.3.1 Technological Progress and Cooperative Sensing

As a result of technological developments and the steadily increasing performance of the information infrastructure, remote sensing has rapidly advanced over the last decade (Toth and Józków, 2016). In particular, developments regarding imaging sensors and new platforms are facilitating numerous applications and, to some extent, they are even widening the range of applications. This can be observed, for instance, in the development of unmanned aerial systems (UASs), which are nowadays increasingly used for the exploration of areas that are difficult to access with terrestrial systems, but also for the acquisition of imagery to serve as the basis for the creation of 3D reconstructions (e.g. of cultural heritage objects or building roofs), for the monitoring of agriculture/farming and ecosystem dynamics, or for the monitoring of hydrodynamics associated with physical and biological processes (Martinsanz, 2016). At the same time as such developments, the performance of the general information infrastructure has improved, mainly affecting the processing, storage and transmission of data. This facilitates the handling of larger amounts of data resulting from the consideration of larger areas, an increasing level of detail, and the use of multi-sensory systems. Despite the use of platforms equipped with multiple sensors, more and more applications are now involving multiple platforms and thus a form of *cooperative sensing* (Toth and Józków, 2016), where the platforms share the acquired (typically multi-sensory) data, for example, to obtain improved geo-localisation or to derive a better reconstruction of the object space. UAS swarms or the emerging micro-satellite systems particularly rely on the concept of such a cooperative sensing.

15.3.2 Spaceborne Platforms for Optical High-resolution Earth Observation

In recent years, a number of new Earth observation platforms have been designed and launched to orbit that have significantly changed the application field of remote sensing. In early 2000, this started with high-resolution optical sensors like IKONOS and QuickBird, opening a wide range of applications in urban remote sensing that could not be addressed by classical optical sensor systems like Landsat. The specifications for the first high-resolution optical satellite sensors faced a trade-off between improvement of resolution in terms of ground sampling distance and providing only a few channels in the visible and near-infrared spectrum, leading to a lack of spectral information. This drawback was overcome by systems like WorldView-2, which covers a larger spectral domain from 400 to 1040 nm with eight channels. Besides providing high-resolution data, WorldView-3 also captures data with larger ground sampling distance in the spectral range from 405 to 2365 nm. These additional channels are mainly in the short infrared, but some are also dedicated to providing observations for atmospheric corrections, like the cirrus band of WorldView-3 CAVIS. This trend also holds for mid-resolution systems. Landsat-8 increased the number of channels compared to Landsat-7 and former systems, namely band 1 (coastal aerosol), two bands with smaller bandwidth in the thermal range instead of just one, and band 9 (cirrus) in the range 1360–1380 nm. Sentinel-2A, launched in 2015, has 12 bands in the spectral range 440–2300 nm, with different ground sampling distances ranging from 10 to 60 m. In common with Landsat-8, its band 10 (ground sampling distance of 60 m) is provided for atmospheric correction, but provides a higher spectral sampling at the red-edge (bands 5, 6 and 7 with ground sampling distance of 20 m) and near-infrared (band 8A and 8 with ground sampling distances of 20 and 10 m, respectively) combined with smaller bandwidths. The improvements in spectral information are also indicated by principal component analysis. Compared to the data of Landsat-7 and former systems, the higher principal components of Sentinel-2 data also carry information. Even more flexibility in terms of multi-temporal data acqusition is provided by the 'Doves' swarm system of Planet Labs. Combined with the fact that Landsat and Sentinel-2 data can be accessed without charge, an even increased use of remote sensing data can be expected.

15.3.3 Crowdsourcing and Crowdsensing

Since data acquisition and annotation are typically tasks that are time-consuming and costly, recent trends towards a *crowdsourcing* can be observed. Although this general term covers a range of activities, the remote sensing and computer vision society mainly benefits from *community-contributed data*, to which great attention has been paid in recent years. There has also been emerging interest in the crowdsourcing of geospatial data by generating a map using informal social networks and Web 2.0 technology, for which a valuable review can be found in (Heipke, 2010). While traditional mapping has mainly been coordinated and conducted by large organisations such as national mapping agencies, mapping via crowsdsourcing potentially involves very large user groups who often collaborate voluntarily and without financial interest. This in turn results in the availability of more and more data sets at very low costs (partially even open-access data sets) and with steady extensions of the mapped area and with frequent updates. The main limitation of such crowdsourcing, however, is a lack of quality assurance, as users might lack formal training in map making. Accordingly, recent investigations also focus on approaches for quality assurance for such data (Goodchild and Li, 2012).

In contrast to mapping purposes, community-contributed data can also be used as source of knowledge about places and about the interests, behaviours and mobility of people if the contents have spatial and temporal references (Andrienko et al., 2009). This is of particular interest for local administrations, such as tourist services or advertising agencies. This data can be accessed in the form of Flickr photos, for example, and be used to analyse concentrations and movements of tourists at the scale of cities or regions. With the availability of both spatial and temporal references, community-contributed photos can be used in two different ways. They can be used as a collection of independent spatio-temporal events that allows us to study properties of places, which in turn is important to find places of interest or extract temporal patterns of visits to these places. In contrast, community-contributed photos with spatial and temporal references can be used by taking into account the individual contributors, which allows us to study the properties and behaviour of people, as photos taken by the same person can be used to construct a trajectory of this person through geographical space, and the whole data set thus corresponds to a set of trajectories of multiple people.

The availability of community-contributed data is also fostered by the fact that our environment is becoming equipped with more and more sensors, for instance mounted on vehicles or infrastructure (Toth and Józków, 2016). Thus, we are creating an increasingly large volume of data – voluntarily or involuntarily – and, together with the data collected by people themselves, crowdsourcing becomes a kind of *crowdsensing* (Ganti et al., 2011). The daily amount of worldwide acquired data reached 2.5 quintillion bytes of data in 2012 (IBM, 2012), and 90% of the data in the world has been acquired in the past two years. Among this collected data, image and video data represent the largest amount, and, of these, a relatively large part is stored on the Web (e.g. on Flickr or YouTube). Certainly, the quality of data acquired via crowdsensing is still not comparable to that acquired by professional organisations, but, due to steady technological advancements, as well as progress in data processing, there may be a point when crowdsensing data achieves a quality that is similar to current mapping technologies (Toth and Józków, 2016).

15.3.4 Large-scale Considerations

Due to the increasingly large volume of acquired data, a scalability of approaches towards data-intensive processing is often required. One thus has to take into account that a massive amount of data typically results from the observation of large areas and dense sampling (i.e. high resolution of the acquired data). In particular, this is a major challenge for applications relying on airborne or satellite imagery for the analysis of *large-scale environments*. In this regard, much effort has been spent on land-cover and land-use classification, for example, based on hyperspectral imagery (Camps-Valls et al., 2014; Keller et al., 2016; Plaza et al., 2009).

Another topic of great interest addresses *large-scale 3D reconstruction* (Frahm et al., 2010; Furukawa et al., 2010), where a huge set of images of a scene is exploited in order to densely reconstruct the 3D structure of a scene. Due to the rapid increase in available imagery and due to technological advances in computing, 3D reconstruction can be performed at a massive scale. Recent approaches even focus on the simultaneous reconstruction of both the 3D structure and a segmentation into semantic object classes, which is referred to as *semantic 3D reconstruction* (Bláha et al., 2016). The same observation can be made for applications involving laser scanning technology and focusing on the direct acquisition of 3D data of districts or cities (Paparoditis et al., 2014; Serna et al., 2014), or even the acquisition of 3D data at the level of a whole nation (Stoter et al., 2016; Vosselman et al., 2015). Consequently, the scalability of approaches towards data-intensive processing is becoming more and more important. This has, for instance, been addressed with recent approaches focusing on the enrichment of acquired data with semantic information (Hackel et al., 2016; Smeeckaert et al., 2013; Serna and Marcotegui, 2014; Weinmann et al., 2015) or the creation of 3D models from the acquired data (Lafarge and Mallet, 2012; Vosselman et al., 2015). Furthermore, it has to be taken into account that, simultaneously to the consideration of larger and larger areas of interest, the level of detail of the acquired data is steadily increasing due to the increasing performance of respective acquisition systems in recent years. For the example of acquiring 3D data for larger urban areas, early approaches mostly relied on the use of airborne laser scanning data with a few points per square metre, while nowadays most of the presented approaches for scene analysis focus on ground-based data acquisition with a point density of hundreds or thousands of points per square metre, depending on the application and the utilised device.

One also has to take into account that the steadily increasing size of data sets is typically accompanied by an increasing number of considered classes, where classes are partially also organised in class hierarchies. The classification task thus faces the challenge of a *large-scale learning*. This is the main focus of a recent benchmark for large-scale image classification and object detection based on large image collections such as the ImageNet database (Deng et al., 2009; Russakovsky et al., 2015) with millions of images and hundreds of object categories. Similarly, large-scale learning is addressed with a benchmark on scene understanding and scene categorisation, which has been addressed based on a data set covering a richness and variety of environmental scenes (Xiao et al., 2016) and, with a particular focus on scene recognition, a significantly larger data set has recently been presented (Zhou et al., 2014).

15.3.5 Benchmarking

For many research domains, it is of great importance that a statement about the performance of the proposed approaches can be made. For this reason, it is desirable to compare the result obtained to some kind of reference. This comparison allows us to judge about the suitability of an approach, and a further comparison with different approaches with respect to the reference allows us to draw conclusions about which approach is suited best to solve a certain problem. While the comparison to a reference has of course been the subject of a huge number of investigations, most of the involved data is not shared for other people to test their own approaches on the same data. This would however be particularly meaningful if one considers that the results of any approach may vary significantly from one data set to another. Accordingly, it is desirable to have publicly available benchmark data sets on which the performance of approaches addressing a certain problem can be evaluated. In the last decade, the number of publicly available benchmark data sets has drastically increased, addressing novel research domains and partially representing more challenging data sets. Among others, a diversity of benchmark data sets have recently been published for multi-platform photogrammetry (Nex et al., 2015), mobile robotics and autonomous driving research (Geiger et al., 2013), 3D scene analysis (Munoz et al., 2009; Rottensteiner et al., 2012; Serna and Marcotegui, 2014; Paparoditis et al., 2014; Hackel et al., 2016), classification of hyperspectral imagery (Braun et al., 2015), material classification (Sharan et al., 2009; Weinmann et al., 2014) and object recognition and image classification (Deng et al., 2009; Russakovsky et al., 2015).

15.3.6 Consideration of Spatial Context

For many applications, it is common to treat every pixel of an image or every point of a point cloud individually as a data point. For a classification task, this corresponds to extracting one feature vector for each data point to be classified. Typical examples following this strategy can for instance be found in the classification of hyperspectral data for land-cover or land-use monitoring, where the intensity values across numerous spectral channels may be used to define a feature vector, or in the classification of 3D point cloud data for scene interpretation, where nowadays typically both geometric and radiometric features may be extracted for each 3D point and sampled as a feature vector. As a consequence of the individual treatment of each data point, however, the visualisation of respective classification results as a classified image or classified point cloud typically reveals noisy behaviour. To address the fact that spatially neighbouring data points (i.e. neighbours on an image raster or neighbours in a local 3D neighbourhood of a point cloud) tend to have correlated class labels, it is advisable to involve information about the local context by considering the class labels and/or the feature vectors of data points within a spatial vicinity. For this purpose, contextual learning techniques may be involved that model relationships within this local vicinity via statistical models. Among these, Markov random fields (MRFs) or conditional random fields (CRFs) represent the most commonly applied approaches.

15.3.7 Multi-scale Representations

A further issue that has to be taken into account is the size of the support regions that are used for calculating features (e.g. for a classification task). More specifically, the size of the support region is typically referred to as the *scale* which may describe the size of a considered image region or the size of a considered local 3D neighbourhood. Whereas features on a smaller scale typically address fine details, features on a larger scale provide a rather coarse description and thus typically introduce a smoothing effect. Consequently, an adequate choice of the scale parameter is of great importance. Depending on the application, it may also be feasible to process features on multiple scales, as this not only allows us to derive features describing the local structure in image regions or 3D neighbourhoods of different scale, but also to describe the behaviour of these features across the different scales (Weinmann, 2016).

15.3.8 Multiple Classifier Systems

While classifier comparisons continue to be one of the "favourite pastimes" of remote sensing, frequently stating which classifiers perform better or best of all compared classifiers (Braun et al., 2015, 2012), it has been very clear for a long time that striving to find the best of all classifiers cannot succeed. As the so-called no-free-lunch theorem points out, a separation function that performs

Table 15.1 Example of a simple MCS (majority voting). Given four different classifiers $C_{1,...,4}$ which assign labels $Y_{C_j} \in [1,2,3]$ to six points $X \in [1,\ldots,6]$, by comparison to the true labels Y_{tr} different accuracy values are produced. Majority voting now simply assigns the label assigned most frequently by $C_{1,...,4}$ as $Y_{MCS} \in [1,2,3]$ and improves accuracy (in an admittedly optimistic fashion).

X	C_1	C_2	C_3	C_4	**MCS**
Point 1 ($Y_{tr} = 1$)	1	1	1	2	1
Point 2 ($Y_{tr} = 1$)	3	1	4	1	1
Point 3 ($Y_{tr} = 2$)	2	1	2	3	2
Point 4 ($Y_{tr} = 2$)	2	2	3	1	2
Point 5 ($Y_{tr} = 3$)	3	3	3	3	3
Point 6 ($Y_{tr} = 3$)	3	3	3	2	3
Accuracy	83.3	83.3	66.7	33.3	100.0

better on a certain separation problem will perform worse on others (Wolpert and Macready, 1997; Wolpert, 2002). Hence, trying to develop "universally optimal classifiers" is in vain. Recognising this, a lot of effort has been invested recently in trying to combine the predictions of several independent classifiers, instead of identifying the best one. Such approaches can be addressed as multiple classifier systems (MCSs). Recent reviews in remote sensing have been presented by Benediktsson et al. (2007), Smits (2002), and Briem et al. (2002), and an international workshop series held over a decade has dedicated itself to discussing MCS – the results are documented in several books e.g (Sansone et al., 2011).

In fact, many MCS approaches are sophisticated and based on complex mathematics. However, even to the less experienced user, straightforward approaches exist to combine the outputs of different classifiers. For instance, using a set of $j = 4$ different classifiers $C_{1,...,4}$, which assign class labels Y_{C_j} to feature vectors X, a simple majority voting (i.e. treating the label most frequently assigned on each pixel as the final label) may improve the results (Table 15.1).

Slightly more complex approaches would combine the posterior probabilities of the individual classifiers, for instance, via product functions. Hence, using MCS to improve one's results is not as complex as it may look at first glance, and we encourage remote sensing users to try to improve their results by applying an MCS, especially as many modern toolboxes, such as the PR-Toolbox by Duin et al. (2000) (now called perClass), provide comprehensive and easy to use MCS functionality.

15.3.9 Augmented Reality

Augmented reality (AR) or *mixed reality* extends a user's view of the real world with virtual computer-generated objects which provide additional information

data. *IEEE Journal of Selected Topics in Applied Earth Observations and Remote Sensing*, 5(2):436–443.

Braun, A., Weinmann, M., Keller, S., Müller, R., Reinartz, P., and Hinz, S. (2015). The EnMAP contest: Developing and comparing classification approaches for the Environmental Mapping and Analysis Programme – Dataset and first results. *The International Archives of the Photogrammetry, Remote Sensing and Spatial Information Sciences*, 40(3/W3):169–175.

Briem, G., Benediktsson, J., and Sveinsson, J. (2002). Multiple classifiers applied to multisource remote sensing data. *IEEE Transactions on Geoscience and Remote Sensing*, 40(10):2291–2299.

Bryant, R., Thoma, D., Moran, S., Goodrich, D., Keefer, T., Paige, G., and Skirvin, S. (2003). Evaluation of hyperspectral, infrared temperature and radar measurements for monitoring surface soil moisture. In *Proceedings of the First Interagency Conference on Research in the Watersheds*, pages 528–533.

Camps-Valls, G., Tuia, D., Bruzzone, L., and Benediktsson, J. (2014). Advances in hyperspectral image classification: Earth monitoring with statistical learning methods. *IEEE Signal Processing Magazine*, 31(1):45–54.

Choy, C., Xu, D., Gwak, J., Chen, K., and Savarese, S. (2016). 3D-R2N2: A unified approach for single and multi-view 3D object reconstruction. *CoRR*, abs/1604.00449: 1–17.

Ciampalini, A., André, F., Garfagnoli, F., Grandjean, G., Lambot, S., Chiarantini, L., and Moretti, S. (2015). Improved estimation of soil clay content by the fusion of remote hyperspectral and proximal geophysical sensing. *Journal of Applied Geophysics*, 116:135–145.

Demotes-Mainard, S., Boumaza, R., Meyer, S., and Cerovic, Z. (2008). Indicators of nitrogen status for ornamental woody plants based on optical measurements of leaf epidermal polyphenol and chlorophyll contents. *Scientia Horticulturae*, 115(4):377–385.

Deng, J., Dong, W., Socher, R., Li, L.-J., Li, K., and Fei-Fei, L. (2009). ImageNet: A large-scale hierarchical image database. In *Proceedings of the IEEE Conference on Computer Vision and Pattern Recognition*, pages 248–255.

Duin, R., Juszczak, P., Paclik, P., Pekalska, E., De Ridder, D., Tax, D., and Verzakov, S. (2000). A matlab toolbox for pattern recognition. *PRTools version* 3.

Fischer, P., Dosovitskiy, A., Ilg, E., Häusser, P., Hazirbas, C., Golkov, V., van der Smagt, P., Cremers, D., and Brox, T. (2015). FlowNet: Learning optical flow with convolutional networks. *CoRR*, abs/1504.06852:1–13.

Frahm, J.-M., Pollefeys, M., Lazebnik, S., Clipp, B., Gallup, D., Raguram, R., and Wu, C. (2010). Fast robust reconstruction of large scale environments. In *Proceedings of the 44th Annual Conference on Information Sciences and Systems*, pages 1–5.

Furukawa, Y., Curless, B., Seitz, S., and Szeliski, R. (2010). Towards internet-scale multi-view stereo. In *Proceedings of the IEEE Conference on Computer Vision and Pattern Recognition*, pages 1434–1441.

Ganti, R., Ye, F., and Lei, H. (2011). Mobile crowdsensing: Current state and future challenges. *IEEE Communications Magazine*, 49(11):32–39.

Geiger, A., Lenz, P., Stiller, C., and Urtasun, R. (2013). Vision meets robotics: The KITTI dataset. *International Journal of Robotics Research*, 32(11):1231–1237.

Goffart, J., Olivier, M., and Frankinet, M. (2008). Potato crop nitrogen status assessment to improve N fertilization management and efficiency: Past – present – future. *Potato Research*, 51:355–383.

Goodchild, M. and Li, L. (2012). Assuring the quality of volunteered geographic information. *Spatial Statistics*, 1:110–120.

Goodfellow, I., Bengio, Y., and Courville, A. (2016). *Deep learning*. MIT Press, Cambridge, MA, USA.

Hackel, T., Wegner, J., and Schindler, K. (2016). Fast semantic segmentation of 3D point clouds with strongly varying density. *ISPRS Annals of the Photogrammetry, Remote Sensing and Spatial Information Sciences*, 3(3):177–184.

Han, X., Leung, T., Jia, Y., Sukthankar, R., and Berg, A. (2015). MatchNet: Unifying feature and metric learning for patch-based matching. In *Proceedings of the IEEE Conference on Computer Vision and Pattern Recognition*, pages 3279–3286.

Hanjalic, A., Lienhart, R., Ma, W., and Smith, J. (2008). The holy grail of multimedia information retrieval: So close or yet so far away? *Proceedings of the IEEE*, 96(4):541–547.

Hassan-Esfahani, L., Torres-Rua, A., Jensen, A., and McKee, M. (2015). Assessment of surface soil moisture using high-resolution multi-spectral imagery and artificial neural networks. *Remote Sensing*, 7(3):2627–2646.

He, K., Zhang, X., Ren, S., and Sun, J. (2016). Deep residual learning for image recognition. In *Proceedings of the IEEE Conference on Computer Vision and Pattern Recognition*, pages 770–778.

Heipke, C. (2010). Crowdsourcing geospatial data. *ISPRS Journal of Photogrammetry and Remote Sensing*, 65(6):550–557.

Homolová, L., Malenovský, Z., Clevers, J., García-Santos, G., and Schaepman, M. (2013). Review of optical-based remote sensing for plant trait mapping. *Ecological Complexity*, 15(8):1–16.

Hübner, P., Weinmann, M., Hillemann, M., Jutzi, B., and Wursthorn, S. (2018). Combining independent visualization and tracking systems for augmented reality. *The International Archives of the Photogrammetry, Remote Sensing and Spatial Information Sciences*, 42(2):455–462.

Hübner, P., Landgraf, S., Weinmann, M., and Wursthorn, S. (2019). Evaluation of the Microsoft HoloLens for the mapping of indoor building environments. In Kersten, T., editor, 39. Wissenschaftlich-Technische Jahrestagung der DGPF, pages 44–53. Deutsche Gesellschaft für Photogrammetrie, Fernerkundung und Geoinformation (DGPF), Munich, Germany.

IBM (2012). What is big data? http://www-01.ibm.com/software/data/bigdata/what-is-big-data.html (accessed 30 August 2016).

Jacquemoud, S., Verhoef, W., Baret, F., Bacour, C., Zarco-Tejada, P., Asner, G., François, C., and Ustin, S. (2009). PROSPECT + SAIL models: A review of use for vegetation characterization. *Remote Sensing of Environment*, 113:S56–S66.

Keller, S., Braun, A., Hinz, S., and Weinmann, M. (2016). Investigation of the impact of dimensionality reduction and feature selection on the classification of hyperspectral EnMAP data. In *Proceedings of the 8th Workshop on Hyperspectral Image and Signal Processing: Evolution in Remote Sensing*, pages 1–5.

Kendall, A., Grimes, M., and Cipolla, R. (2015). PoseNet: A convolutional network for real-time 6-DOF camera relocalization. In *Proceedings of the IEEE International Conference on Computer Vision*, pages 2938–2946.

Kerl, C., Sturm, J., and Cremers, D. (2013). Robust odometry estimation for RGB-D cameras. In *Proceedings of the IEEE International Conference on Robotics and Automation*, pages 3748–3754.

Krizhevsky, A., Sutskever, I., and Hinton, G. (2012). ImageNet classification with deep convolutional neural networks. *Advances in Neural Information Processing Systems*, pages 1097–1105.

Lafarge, F. and Mallet, C. (2012). Creating large-scale city models from 3D-point clouds: A robust approach with hybrid representation. *International Journal of Computer Vision*, 99(1):69–85.

LeCun, Y., Bengio, Y., and Hinton, G. (2015). Deep learning. *Nature*, 7553(521):436–444.

Li, F., Gnyp, M., Jia, L., Miao, Y., Yu, Z., Koppe, W., Bareth, G., Chen, X., and Zhang, F. (2008). Estimating N status of winter wheat using a handheld spectrometer in the North China Plain. *Field Crops Research*, 106(1):77–85.

Lin, F., Qiu, L., Deng, J., Shi, Y., Chen, L., and Wang, K. (2010). Investigation of SPAD meter-based indices for estimating rice nitrogen status. *Computers and Electronics in Agriculture*, 71:S60–S65.

Long, J., Shelhamer, E., and Darrell, T. (2015). Fully convolutional networks for semantic segmentation. In *Proceedings of the IEEE Conference on Computer Vision and Pattern Recognition*, pages 3431–3440.

Martinsanz, G. (2016). Unmanned aerial vehicles in geomatics. *International Journal of Geo-Information*, 5(8):147–148.

Muñoz-Huerta, R., Guevara-Gonzalez, R., Contreras-Medina, L., Torres-Pacheco, I., Prado-Olivarez, J., and Ocampo-Velazquez, R. (2013). A review of methods for sensing the nitrogen status in plants: Advantages, disadvantages and recent advances. *Sensors*, 13(8):10823–10843.

Munoz, D., Bagnell, J., Vandapel, N., and Hebert, M. (2009). Contextual classification with functional max-margin Markov networks. In *Proceedings of the IEEE Conference on Computer Vision and Pattern Recognition*, pages 975–982.

Nex, F., Gerke, M., Remondino, F., Przybilla, H.-J., Bäumker, M., and Zurhorst, A. (2015). ISPRS benchmark for multi-platform photogrammetry. *ISPRS Annals of the Photogrammetry, Remote Sensing and Spatial Information Sciences*, 2(3/W4):135–142.

Nissen, H. (1986). The archaic texts from Uruk. *World Archaeology*, 17(3):317–334.

Paparoditis, N., Vallet, B., Marcotegui, B., and Serna, A. (2014). IQmulus & TerraMobilita Contest – Analysis of mobile laser scans (MLS) in dense urban environments. MATIS Laboratory, French National Mapping Agency (IGN) and Center for Mathematical Morphology (CMM), MINES ParisTech, http://data.ign.fr/benchmarks/UrbanAnalysis/ (accessed 30 August 2014).

Plaza, A., Benediktsson, J., Boardman, J., Brazile, J., Bruzzone, L., Camps-Valls, G., Chanussot, J., Fauvel, M., Gamba, P., Gualtieri, A., Marconcini, M., Tilton, J., and Trianni, G. (2009). Recent advances in techniques for hyperspectral image processing. *Remote Sensing of Environment*, 113:S110–S122.

Roscher, R., Bohn, B., Duarte, M. F. and Garcke, J. (2019). Explainable machine learning for scientific insights and discoveries, available at: https://arxiv.org/abs/1905.088832019 (last access Aug. 25, 2019)

Rottensteiner, F., Sohn, G., Jung, J., Gerke, M., Baillard, C., Benitez, S., and Breitkopf, U. (2012). The ISPRS benchmark on urban object classification and 3D building reconstruction. *ISPRS Annals of the Photogrammetry, Remote Sensing and Spatial Information Sciences*, 1(3):293–298.

Russakovsky, O., Deng, J., Su, H., Krause, J., Satheesh, S., Ma, S., Huang, Z., Karpathy, A., Khosla, A., Bernstein, M., Berg, A., and Fei-Fei, L. (2015). ImageNet large scale visual recognition challenge. *International Journal of Computer Vision*, 115(3):211–252.

Sansone, C., Kittler, J., and Roli, F. (2011). *Multiple Classifier Systems: 10th International Workshop, MCS 2011, Naples, Italy, June 15-17, 2011. Proceedings*, Vol. 6713. Springer Science & Business Media, Berlin, Germany.

Schmidhuber, J. (2015). Deep learning in neural networks: An overview. *Neural Networks*, 61:85–117.

Sermanet, P., Kavukcuoglu, K., Chintala, S., and LeCun, Y. (2013). Pedestrian detection with unsupervised multi-stage feature learning. In *Proceedings of the IEEE Conference on Computer Vision and Pattern Recognition*, pages 3626–3633.

Serna, A. and Marcotegui, B. (2014). Detection, segmentation and classification of 3D urban objects using mathematical morphology and supervised learning. *ISPRS Journal of Photogrammetry and Remote Sensing*, 93:243–255.

Serna, A., Marcotegui, B., Goulette, F., and Deschaud, J.-E. (2014). Paris-rue-Madame database: A 3D mobile laser scanner dataset for benchmarking urban detection, segmentation and classification methods. In *Proceedings of the International Conference on Pattern Recognition Applications and Methods*, pages 819–824.

Sharan, L., Rosenholtz, R., and Adelson, E. (2009). Material perception: What can you see in a brief glance? *Journal of Vision*, 9(8):784.

Shaver, T., Khosla, R., and Westfall, D. (2011). Evaluation of two crop canopy sensors for nitrogen variability determination in irrigated maize. *Precision Agriculture*, 12(6):892–904.

Shen, L., Wee Chua, T., and Leman, K. (2015). Shadow optimization from structured deep edge detection. In *Proceedings of the IEEE Conference on Computer Vision and Pattern Recognition*, pages 2067–2074.

Sleeswyk, A. (1982). The prow of the 'Nike of Samothrace' reconsidered. *International Journal of Nautical Archaeology*, 11(3):233–243.

Smeeckaert, J., Mallet, C., David, N., Chehata, N., and Ferraz, A. (2013). Large-scale classification of water areas using airborne topographic lidar data. *Remote Sensing of Environment*, 138:134–148.

Smith, R. (1983). Nicephore Niepce in England. *History of Photography*, 7(1):43–50.

Smits, P. (2002). Multiple classifier systems for supervised remote sensing image classification based on dynamic classifier selection. *IEEE Transactions on Geoscience and Remote Sensing*, 40(4):801–813.

Stoter, J., Vallet, B., Lithen, T., Pla, M., Wozniak, P., Kellenberger, T., Streilein, A., Ilves, R., and Ledoux, H. (2016). State-of-the-art of 3D national mapping in 2016. *The International Archives of the Photogrammetry, Remote Sensing and Spatial Information Sciences*, 41(B4):653–660.

Szegedy, C., Liu, W., Jia, Y., Sermanet, P., Reed, S., Anguelov, D., Erhan, D., Vanhoucke, V., and Rabinovich, A. (2015). Going deeper with convolutions. In *Proceedings of the IEEE Conference on Computer Vision and Pattern Recognition*, pages 1–9.

Szeliski, R. (2010). *Computer vision: Algorithms and applications*. Springer, Berlin, Germany.

Taylor, G., Fergus, R., LeCun, Y., and Bregler, C. (2010). Convolutional learning of spatio-temporal features. In *Proceedings of the European Conference on Computer Vision*, pages 140–153.

Toth, C. and Jóżków, G. (2016). Remote sensing platforms and sensors: A survey. *ISPRS Journal of Photogrammetry and Remote Sensing*, 115:22–36.

Urban, S., Leitloff, J., Wursthorn, S., and Hinz, S. (2013). Self-localization of a multi-fisheye camera based augmented reality system in textureless 3D building models. *ISPRS Annals of the Photogrammetry, Remote Sensing and Spatial Information Sciences*, 2(3/W2):43–48.

Urban, S., Leitloff, J., and Hinz, S. (2015). Improved wide-angle, fisheye and omnidirectional camera calibration. *ISPRS Journal of Photogrammetry and Remote Sensing*, 108:72–79.

Urban, S., Weinmann, M., and Hinz, S. (2017a). mdBRIEF - A fast online-adaptable, distorted binary descriptor for real-time applications using calibrated wide-angle or fisheye cameras. *Computer Vision and Image Understanding*, 162:71–86.

Urban, S., Wursthorn, S., Leitloff, J., and Hinz, S. (2017b). MultiCol bundle adjustment: A generic method for pose estimation, simultaneous self-calibration and reconstruction for arbitrary multi-camera systems. *International Journal of Computer Vision*, 121(2):234–252.

Valladas, H., Cachier, H., Maurice, P., de Quirost, F., Clottes, J., Valdes, V., Uzquiano, P., and Arnold, M. (1992). Direct radiocarbon dates for prehistoric paintings at the Altamira, El Castillo and Niaux caves. *Nature*, 357(1992):68–70.

Vosselman, G., Oude Elberink, S., Post, M., Stoter, J., and Xiong, B. (2015). From nationwide point clouds to nationwide 3D landscape models. In *Proceedings of the Photogrammetric Week 2015*, pages 247–256.

Weinmann, M. (2016). *Reconstruction and analysis of 3D scenes: From irregularly distributed 3D points to object classes*. Springer, Cham, Switzerland.

Weinmann, M., Gall, J., and Klein, R. (2014). Material classification based on training data synthesized using a BTF database. In *Proceedings of the European Conference on Computer Vision*, pages 156–171.

Weinmann, M., Urban, S., Hinz, S., Jutzi, B., and Mallet, C. (2015). Distinctive 2D and 3D features for automated large-scale scene analysis in urban areas. *Computers & Graphics*, 49:47–57.

Wohlhart, P. and Lepetit, V. (2015). Learning descriptors for object recognition and 3D pose estimation. In *Proceedings of the IEEE Conference on Computer Vision and Pattern Recognition*, pages 3109–3118.

Wolpert, D. and Macready, W. (1997). No free lunch theorems for optimization. *IEEE Transactions on Evolutionary Computation*, 1(1):67–82.

Wolpert, D. H. (2002). The supervised learning no-free-lunch theorems. In *Soft Computing and Industry*, pages 25–42. Springer, New York, USA.

Wursthorn, S., Coelho, A., and Staub, G. (2004). Applications for mixed reality. *The International Archives of the Photogrammetry, Remote Sensing and Spatial Information Sciences*, 35(B3):1049–1054.

Xiao, J., Ehinger, K., Hays, J., Torralba, A., and Oliva, A. (2016). SUN database: Exploring a large collection of scene categories. *International Journal of Computer Vision*, 119(1):3–22.

Younis, S. and Iqbal, J. (2015). Estimation of soil moisture using multispectral and FTIR techniques. *The Egyptian Journal of Remote Sensing and Space Science*, 18(2):151–161.

Zagoruyko, S. and Komodakis, N. (2016). Wide residual networks. *CoRR*, abs/1605.07146:1–13.

Zbontar, J. and LeCun, Y. (2015). Computing the stereo matching cost with a convolutional neural network. In *Proceedings of the IEEE Conference on Computer Vision and Pattern Recognition*, pages 1592–1599.

Zhao, K., Valle, D., Popescu, S., Zhang, X., and Mallick, B. (2013). Hyperspectral remote sensing of plant biochemistry using Bayesian model averaging with variable and band selection. *Remote Sensing of Environment*, 132:102–119.

Zhou, B., Lapedriza, A., Xiao, J., Torralba, A., and Oliva, A. (2014). Learning deep features for scene recognition using places database. *Advances in Neural Information Processing Systems*, 27:487–495.

Note

1. Of course, this argument may provoke fervent debate on whether "complexity" should be defined when talking about images. We simply hope that the reader benevolently embraces what we are trying to say here.

Index